THE
UNDERSTANDING
OF ANIMALS

NEW SCIENTIST GUIDES
Series Editor: *Colin Tudge*

The Understanding of Animals
Edited by *Georgina Ferry*

Observing the Universe
Edited by *Nigel Henbest*

Grow Your Own Energy
Edited by *Mike Cross*

Building the Universe
Edited by *Christine Sutton*

The Making of the Earth
Edited by *Richard Fifield*

A new**scientist** GUIDE

THE
UNDERSTANDING
OF ANIMALS

Edited by
GEORGINA FERRY

Basil Blackwell & New Scientist

© Articles and editorial, IPC Magazines Ltd.
Volume rights, Basil Blackwell Limited.

First published in book form in 1984 by
Basil Blackwell Limited.
108 Cowley Road, Oxford OX4 1JF.

British Library Cataloguing in Publication Data

The understanding of animals. – (New scientist
  guides)
1. Zoology
I. Ferry, Georgina
591   QL45.2
ISBN 0-85520-729-9
ISBN 0-85520-728-0 Pbk

Typeset by Oxford Verbatim Limited
Printed and bound in Great Britain by
Bell and Bain Ltd., Glasgow

"The welfare of animals must depend on an *understanding* of animals, and one does not come by this understanding intuitively: it must be learned."

Sir Peter Medawar

Contents

Contents

Contents   ix

List of Illustrations

Frontispiece, Pete Addis: p. 10 Frieder Sauer, Bruce Coleman: p. 36 Sally Anne Thompson, Animal Photographers: p. 40 Diana and Nick Sullivan, Bruce Coleman: p. 44 Richard Rawlins: p. 46 Carol Berman: p. 51, 55, 56 Tim Clutton-Brock: p. 62, 65 Eric Hosking: p. 72–73 Marconi Research Centre: p. 78 David MacDonald: p. 93 Douglas Rhodes: p. 96 Chris Jones: p. 100 Bruce Coleman: p. 101 Richard Vaughan, ARDEA: p. 111, 116 A. H. Harcourt, Cambridge: p. 112, 113 M. Simpson, Cambridge: p. 120 Kenneth W. Fink, ARDEA: p. 121 ARDEA: p. 123 ARDEA: p. 125 L. R. Dawson, Bruce Coleman: p. 133 Eric Hosking: p. 142 Eric Hosking: p. 153 Radio Times Picture Library, Dimitri Kasterine: p. 156 Bill Potter, Camera Press: p. 162 Lee Lyon, Bruce Coleman: p. 176 Jeff Foott, Bruce Coleman; Eric Hosking; Cynthia Moss; Nova Pic: p. 184 S. Roberts, ARDEA: p. 202 David and Kate Urry, ARDEA: p. 207 'Mr. C. V. A. Peel's little bag', Illustrated London News, *The Sketch* 16 November, 1898: p. 208 Peter Davey, Bruce Coleman: p. 212 Lawrence Mynott: p. 219 Illustration by Ellis Nadler: p. 237 Mary Evans Picture Library: p. 247, 249, 252 Emil Menzel: p. 275 Derek Washington, Bruce Coleman: p. 272 Ian Beams, ARDEA: p. 281 ARDEA: p. 285 Pete Addis, Jim Byrne: p. 292 Kenneth W. Fink, ARDEA: p. 294 Pete Addis: p. 300 Keystone.

Contributors

CHARLES AMLANER is in the Department of Zoology at Walla Walla University, Washington State, USA.

W. S. ANTHONY was a member of the Institute of Experimental Psychology at the University of Oxford. He now lectures in the School of Education at the University of Leicester.

ANTHONY ARAK is a member of the Sub-Department of Animal Behaviour at the University of Cambridge.

TIM BEARDSLEY is a member of the Animal Behaviour Research Group in the Department of Zoology at the University of Oxford.

CLIVE CATCHPOLE is a lecturer in Zoology at Bedford College, University of London. He is the author of *Vocal Communication in Birds* (Edward Arnold, 1979).

M. R. A. CHANCE is Reader in Ethology at the University of Birmingham.

JEREMY CHERFAS is Life Sciences consultant of *New Scientist* and a former demonstrator in the Department of Zoology at the University of Oxford. He is the author of *Man Made Life* (Basil Blackwell, 1982).

R. B. CLARK is Professor of Zoology at the University of Newcastle-upon-Tyne.

TIM CLUTTON-BROCK is a member of the Large Animal Research Group in the Zoology Department at the University of Cambridge.

D. R. CROCKER is at the Worplesdon Laboratory of the Ministry of Agriculture, Fisheries and Food, and is a member of the Sociobiology Group of the British Society for Social Responsibility in Science.

MARIAN STAMP DAWKINS is a lecturer in the Department of Zoology at the University of Oxford.

ROBIN DUNBAR is Associate Professor at the Zoological Institute at the University of Stockholm.

The late SIR ERIC EASTWOOD was author of *Wireless Telegraphy* (Applied Science, 1974) and was Director of Research at Baddow Research Laboratories, Marconi's Wireless Telegraph Co Ltd.

GEORGINA FERRY, formerly science news editor of *New Scientist*, is now its psychology consultant.

CAROL G. GOULD is a research associate in the Department of Biology at the University of Princeton, New Jersey, USA.

JAMES L. GOULD is Associate Professor of Biology at the University of Princeton, New Jersey, USA.

PAUL GREENWOOD lectures in Biology at the University of Durham.

ALEXANDER HARCOURT is studying gorillas at the Karisoke Research Centre, Rwanda.

PAUL HARVEY lectures in Biology at the University of Sussex.

NICHOLAS HUMPHREY is a writer and broadcaster, and a former assistant director of the Sub-Department of Animal Behaviour at the University of Cambridge.

DONALD JENNI is a member of the Department of Zoology at the University of Montana, USA.

JOHN KREBS is a member of the Edward Grey Institute of Field Ornithology at the University of Oxford.

HANS KRUUK is at the Institute of Terrestrial Ecology, Banchory, Kincardineshire.

The late DAVID LACK FRS was a reader in Ornithology at the Edward Grey Institute of Field Ornithology at the University of Oxford.

ROGER LEWIN is a writer on the magazine, *Science* and a former features editor of *New Scientist*.

DAVID MACDONALD is a member of the Animal Behaviour Research Group at the University of Oxford.

DAVID MCFARLAND is Reader in Animal Behaviour at the University of Oxford.

NICHOLAS MACKINTOSH is Head of the Department of Psychology at the University of Cambridge.

EMIL MENZEL is a research psychologist at the State University of New York at Stony Brook, USA.

RICHARD RAWLINS is Scientist in Charge at the Caribbean Primate Research Center.

KELLY J. STEWART is studying gorillas at the Karisoke Research Centre, Kwanda.

HERBERT TERRACE is Professor of Psychology at Columbia University, New York, USA.

GAIL VINES is Biological Sciences Editor of *New Scientist*.

EDWARD O. WILSON is Professor of Zoology at Harvard University, USA.

KEN YASUKAWA is a researcher in the Department of Biology at Beliot College, Wisconsin, USA.

Foreword

New Scientist was born on 22 November 1956 and, with only a few interruptions, has appeared every week since. It was intended as a "popular" magazine about science, and so it has remained; but as one of that élite among popular magazines that also serves the specialists. It has brought together all who are concerned with what science has to say about the world: scientists, technologists, politicians, industrialists, teachers, and a new breed of science writers, some of whom are journalists who found that the ideas of science were the most exciting of all, and some of whom are ex-scientists, who found that commentating was a fine and proper way to indulge their interest.

New Scientist Guides are collations of articles taken from the magazine, sometimes abbreviated a little to avoid repetition, but otherwise presented as they first appeared. The individual pieces have been brought up to date and put in perspective by linking passages, to give a multi-faceted, modern overview of the subject in hand.

Thus we intend the Guides to fulfil two functions. First they describe the main lines of thought and the key events in a given area of science or technology in the past few, crucial decades; but they also show how those events were brought about, how the ideas were first framed, and how they were first perceived by the world at large.

By presenting science and technology in this way the Guides illustrate what is perhaps the greatest scientific discovery of all: that the traditional view of science as an inexorable progression to unequivocal truth is misguided, and that science is, in fact and in essence a product of human thought and human imagination.

Articles appear in *New Scientist* in various guises. Some – roughly distinguishable as those over 1000 words – are acknowledged to be "features", and are printed with the name of their authors: so they appear in the Guides, together with their date of publication.

Shorter pieces tend to be placed in one or other of the magazine's special sections, variously titled "This Week", "Monitor", "Technology", "Comment", "Forum" – and, in the old days, "Notes and Comments". Most of these short articles are published anonymously, which again is how they appear in the Guides, together with their date and the name of the section in which they appeared.

Linking passages, printed in sanserif, have been written specially for the Guides by the Guide editors, and did not appear first in the magazine.

Colin Tudge
Series Editor

Introduction

Humans have taken a lively interest in the doings of their fellow species since their earliest origins, when their chief preoccupations included finding animals to eat and avoiding animals that might eat them. The human species achieved its paramount status only by outwitting all the others; it hit the evolutionary jackpot by acquiring a large brain and the capacity for complex social exchanges (not necessarily in that order). Ever since, *Homo sapiens* has been in a class of its own, even by comparison with its closest relatives. In the competition for resources, it was the inevitable winner, with the power to kill or to domesticate as it saw fit. That power stemmed from its superior understanding, that enabled it to observe animals and to make accurate predictions about how they would behave under different circumstances.

Throughout most of history, our understanding of animals has tended to be limited by our purpose in studying them. Both hunters and farmers have added immeasurably to the existing body of knowledge concerning animal behaviour, but each has sought only to predict behaviour relevant to their own occupation. Beginning with Aristotle in the 4th century BC, philosophers of every age have turned their attention to animals and their habits; their goal was to establish man's place in the natural order. Their first-hand observations tended to be scanty and their interpretations correspondingly extensive; their conclusions depended on their philosophical outlook, animals being regarded sometimes as "brute beasts that have no understanding" and sometimes as superior to humans in their innocence and simplicity.

The 17th century French philosopher René Descartes denied that animals had reason; for him the concept of reason was virtually synonymous with that of the soul, and only humans could be credited with a soul. He also believed that language was a prerequisite for rationality and, as he saw no evidence of language in non-human species, he regarded all animal behaviour as mechanical reflexes. His ideas are still very influential.

The 19th century produced some of the greatest naturalists, if such is an appropriate term for biologists of the stature of Charles Darwin and Alfred Russell Wallace. But it was also the age of the gentleman explorer and the amateur naturalist. These Victorian ladies and gentlemen tended to regard animals as differing only quantitatively from humans in their capacity to think and feel. Unlike the ancients, Victorian naturalists had no need to ask the questions about man's place in nature. They knew that their place at the top of the tree was given by God and, in typically egocentric fashion, tended to interpret the observations they collected so assiduously with reference to their own experience. Festooned with collecting tins and encumbered with anthropomorphic preconceptions, the more dedicated naturalists of those days nevertheless laid the foundations of the modern science of ethology.

The 20th century has seen the goal of understanding what makes animals tick tackled from a wide variety of standpoints. Adherents of the different schools are not as widely divided as they once were, but we are still far from a unified theory of animal behaviour. 20th-century scientists are no more free of preconceptions than any of their predecessors, though they may be able to justify them more convincingly. One of the most influential lines of inquiry began early this century with the experiments of the American psychologists E. L. Thorndike, J. B. Watson and B. F. Skinner. They were disillusioned with introspective approaches to human and animal psychology, distrusted accounts of "instinct" based on anecdotal evidence, and were impressed with the work of the Russian physiologist I. P. Pavlov on conditioned reflexes. Accordingly they confined their subjects (usually rats or pigeons) to laboratories and studied their performance on simple tasks under strictly controlled conditions. They accepted as valid data only such actions as they observed the animals to perform and were, therefore, known as "behaviourists". Part I explores some of the limitations of this approach.

The behaviourists specifically excluded any consideration of the "state of mind" of the animal. They were interested only in documenting the responses animals made when faced with different stimuli and the ways in which these responses could be manipulated, or "conditioned". Inevitably, they came to regard animals as lacking in mental life; they thought all motor learning could be reduced to associations between stimuli and responses. It has taken millions of hours of lever-pressing by countless albino rats in experimental chambers (called Skinner boxes) before the majority of psychologists could be persuaded that this idea was perhaps a little oversimplistic. But the techniques are still very much with us. To this day, rats run mazes and press levers, pigeons peck keys, and psychologists manipulate reward schedules in laboratories all over the world.

Meanwhile, the spiritual descendants of the Victorian naturalists, field zoologists such as Konrad Lorenz in Austria, Niko Tinbergen in Holland and William Thorpe in Cambridge, England, found behaviourism deeply unsatisfying as a route to understanding the behaviour of animals in the wild. It was absurd to generalize from rats and pigeons to the whole of nature and the inside of a Skinner box bore no relation whatever to the natural habitat of any animal. Instead they embraced ethology, the study of animals in their natural surroundings. Their classic studies on sticklebacks, herring gulls, greylag geese and chaffinches (among others) showed that many animals came equipped with innate mechanisms to guide their behaviour – instinct. Preoccupied with the question of what *caused* behaviour, the ethologists and the behaviourists engaged for many years in acrimonious controversy as to whether instinct or learning was more important.

After the Second World War, ethologists became caught up in the resurgence of interest in Charles Darwin's theory of evolution that followed the cracking of the genetic code. Instead of asking what caused behaviour, they began to ask what it was for – in other words, to look for functional explanations of their observations, as opposed to causative ones. They reasoned that natural selection would eliminate any behaviour that was not "adaptive" – that is, that would not increase the likelihood that an individual's genes would be passed on to future generations – and therefore that all observed behaviour must in the long run be adaptive. What is adaptive for one animal may well be maladaptive for another; animals show enormous diversity in the way they choose to live, and the key to this diversity must lie in the diversity of the environments that are available to them. Why else should two very similar species organize their societies in entirely different ways? As illustrated in Parts Two and Three, questions like this drive enthusiastic doctoral students to spend hour after uncomfortable hour observing elusive beasts in inaccessible corners of the world in order to amass enough data on feeding and breeding habits, parental care, communication and every other aspect of behaviour to generate plausible hypotheses.

An animal's environment is made up of a complex set of variables and it may not at first be obvious which are the factors that determine a particular animal's strategy. The computer age now helps ethologists to apply mathematics to evaluate the various influences and come up with a prediction as to how the animal "ought" to behave. Needless to say, a program is only as good as the theory that gave rise to it; it is in the development of theories that ethologists are making their most exciting advances today, as Part Four shows. Ethology now combines with the fields of genetics and energetics in order to put a price on decisions such as when and where to look for food, which mate to choose, how much effort to put into raising young, when to fight and so on. Animal behaviour in the 1980s is fast becoming an exact science.

Those who have tried to extend the hard-line functional approach to human behaviour have encountered intense hostility that would be puzzling if the species were any but our own. Part Five demonstrates that the so-called sociobiologists of the 1970s were not the first to draw analogies between human and animal behaviour. Ever since Charles Darwin, biologists have wondered whether the origins of human behaviour could be found among present-day animal communities. It has been earnestly hoped that by studying the models of harmonious living offered by lesser creatures, sinful humans might find a way to return to such an idyllic existence. Sociobiologists, led by Edward O. Wilson, are not so romantic. They look for universals in the behaviour of humans and other animals and argue that such universal patterns of behaviour must have a strong genetic basis. Their opponents find such genetic determinism highly objectionable when applied to the human species. For them, it implies that humanity is powerless to correct some of its less desirable characteristics. More seriously for them, it provides apparently scientific support for the prevailing political systems of the world, which they see as unjust. This clash of views provides a classic example of how our observations are coloured by our preconceptions. Holding the distorting mirror of politics up to nature, we see what we expect to see.

But to return to our animals: are functional explanations of behaviour as mechanistic in their way as those of the behaviourists? Like behaviourism, the functional approach to animal behaviour has no need to consider the mental life of its subjects and, in their scientific writing at least, ethologists are careful to avoid anthropomorphic interpretations of what they observe. But there are others for whom it is important to ask whether animals are conscious or rational. Psychologists and palaeoanthropologists interested in the evolution of the human intellect look for its origins in subhuman species. Moral philosophers seeking to define how far we should go in exploiting animals need first to assess what constitutes suffering in an animal, beyond obvious physical pain. The flights of fancy of some 19th-century naturalists sowed a deep-rooted mistrust of anthropomorphism that persists to the present day among the scientifically respectable. Many biologists hold (with Descartes) that it is rationality that divides humans from their fellow creatures; to allow other animals even an element of reason or consciousness requires them to revise their assumptions to an unacceptable degree.

But scientists and philosophers now argue persuasively, with experimental evidence to support them, that our dumb friends are not as dumb as they look. Part Six offers several examples. As well as possessing attributes long thought to be exclusive to humans, such as the ability to form representations of objects or concepts that are not physically present, some species possess abilities that are almost beyond our comprehension, such as unerring navigational skill. Although the apparently remarkable capacity of chimpanzees to use linguistic forms of communication (like sign language) turned out to be less remarkable than it first appeared, the experiments revealed potentials in their subjects that would be quite unsuspected by observers of wild populations. Clearly, the animal mind is at last beginning to win a grudging respect from the scientific community.

There is as yet no unified science of animal behaviour but there is a kind of logic in the way the different branches of the subject have evolved in relation to one another. This collection of articles from *New Scientist*, culled from issues covering more than a quarter of a century, aims to illustrate that logic. It does not pretend to be a comprehensive treatise on the behaviour of animals but it does seek to document the diversity of thinking about the origins and purpose of such behaviour. Are animals mindless machines or intelligent (if inscrutable) companions to the human species? Is the prayer book wrong in assuming that they have no understanding? Has our own understanding of them advanced as science has taken an increasingly firm grip on the way we observe them? The articles in this guide demonstrate that the animal behaviour community (if such a thing exists) is far from agreed on these points. But they also reveal the vigour with which the answers to such questions are being pursued.

PART ONE

Animals in the Laboratory:
The Facts and ...

Animals in the Laboratory: The Behaviourists

Studying animals in their natural habitats is not easy. It takes many hours of patient observation, often under very difficult circumstances, before you can say that in environment X, animal Y will perform behaviour Z, 90 per cent of the time. Even then you cannot be sure that your interpretation of what happened was correct. So the possibility of simplifying the whole set-up and controlling it at will must seem very attractive. This was the route taken by the behaviourists in the early decades of the 20th century. They were particularly interested in how animals learn, and concerned to demonstrate their firm belief that their subjects could not solve problems by reasoning, but only through trial and error. Eventually, endless pairings of a behavioural act (first produced by chance) with a desirable outcome would establish the appropriate response. Once acquired, such "conditioned reflexes" were thought to be very hard to extinguish, like well-ingrained habits. Whatever the difficulty of the problem, the animal that solved it was not credited with "cleverness" in having "worked out" the solution; it was regarded as a machine to be switched on or off according to the experimenter's will.

As a route to a general theory of animal behaviour, this technique has limitations that now seem obvious. The behaviourists were also almost certainly mistaken about what was really going on in their own experiments. But laboratory studies of animal learning have not all been a waste of time. R. B. Clark (p. 9) describes attempts to assess the learning capacity of a very simple animal, the ragworm *Nereis*. Although it makes the laboratory rat look like a genius, the ragworm can, with patience, be taught simple avoidance and food-seeking tasks. But is this all there is to the ragworm? The author offers the tantalizing possibility that this animal, with the most rudimentary of nervous systems, might be able to recognise other individual worms.

By the 1950s, psychologists were no longer prepared to take it as read that animals could not reason. The question was, how could the psychologist design a test that would challenge the reasoning capacity of animals but could not be explained away in terms of conditioning theory? W. S. Anthony's article of 1959 (p. 15) looks for evidence of reasoning in the rat and fails to come to any definite conclusion, other than to say that the rat is "brighter than would be expected from simple conditioned-reflex theory".

Twenty years later, experiments were proliferating that challenged the most basic assumptions of the behaviourists. An elegant study (p. 20) suggests that rats form representations of the events which they have learned to associate, and these may be quite different from what we expect. Furthermore, they do have some idea of the consequences of their actions, as Nicholas Mackintosh points out (p. 22). He reassesses conditioning theory in the light of some of the new research.

1

How much can a ragworm learn?

R. B. CLARK

9 March 1961

Ragworms can learn a conditioned response – crawling towards a light that signals food – although they are remarkably slow at learning to avoid an electric shock. How then can one explain their apparent ability to recognise other individual ragworms?

Polychaete worms are the marine relatives of earthworms and include such old friends of the angler as the lugworm *Arenicola* and the ragworm *Nereis*. Although earthworms have been a favourite subject for psychologists, much less attention has been paid to the behaviour of polychaetes, probably because it is so much more convenient to maintain a stock of earthworms in the laboratory than aquaria of marine worms, most of which do not live well away from the sea. Earthworms possess modest learning abilities, but there is reason to suppose that some, at least, of the polychaetes might possess more elaborate behaviour and learn faster than earthworms. A fair proportion of polychaetes live a sedentary existence, but those like *Nereis* that are more active generally have more complicated sense organs and a more complicated brain than any earthworm. They are exposed to a great variety of environmental stimuli and are likely on this account to possess correspondingly flexible behaviour.

Habituation is generally recognised as the simplest type of learning. It consists of a mere dropping out of reactions to repeated, innocuous stimuli. Most animals including even some protozoans (single-celled organisms), manifest it and habituation probably forms a part of the maturing process of all animals. Initially, they react to all sudden stimuli but quickly learn to ignore those which do not have harmful or unpleasant concomitants. Worms that live in

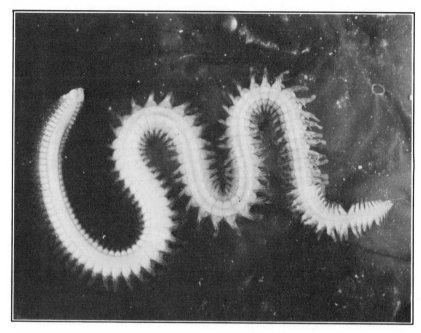

The ragworm Nereis: *a little learning is a dangerous thing*

burrows and emerge at the entrance for feeding, retract violently into the burrow again when startled by such stimuli as the passage of a shadow across them, or by a sudden contact with an object. But time spent retracted inside the burrow is time wasted so far as feeding is concerned and there is considerable biological advantage to any mechanism that reduces unnecessary withdrawals to a minimum. One difficulty for the worm is that it has not sufficiently good eyes to discriminate shapes, and a moving shadow might sometimes presage the approach of a predator, but at other times be caused by nothing more serious than fronds of seaweed washing backwards and forwards in the water. In other words, the same stimulus may indicate different situations on different occasions.

Naturally, one type of stimulus may be more likely to be caused by a predator than another and still others may be totally without significance in the environment in which the animal normally lives. Generally, the greater the biological significance of a stimulus, the more likely the worm is to react to it and the more times it has to be given before the worm becomes habituated and ceases to react at all.

As an example of this, most worms, particularly those living in shallow water, do not react to a sudden increase in light intensity but do react to a sudden decrease. The ragworm, on the other hand, living inter-tidally where it is liable to be exposed by birds turning over stones on the beach, reacts to both sudden increases and decreases in light intensity and habituates more slowly to the former than to the latter stimulus. Dr J. A. C. Nicol at Plymouth, working on the fanworm *Branchiomma*, found that it habituates quite rapidly to a sudden decrease in light intensity but more slowly to a moving shadow, although the change in light intensity may be the same in both cases. The same is true of *Nereis*. Since attack by a predator must always involve movement as well as a possible decrease in the light intensity falling on the worm, the second stimulus has greater biological implication than the first.

Habituation appears to be a rather complicated process in the ragworm *Nereis pelagica*. Crawling over boulders and among sea-weeds, and living in crevices in stones or under the attachment holdfasts of the weed in the inter-tidal zones of rocky shores, it has a very varied and changeable environment. It is attacked and eaten by fishes and crabs when covered at high tide and by shore birds when the beach is exposed at low water. It is continually barraged by stimuli such as moving shadows and mechanical shocks, all of which might indicate the approach of a predator and, under these conditions, one might expect that it would be constantly on the alert and not habituate readily to any repeated stimuli. Nevertheless, it has to emerge from the crevices of the rocks in order to feed and in laboratory experiments we find that it habituates rapidly to a variety of stimuli. In fact, about a third of the worms fail to respond even to the first application of a stimulus and, with very few exceptions, the worms cease to react after two or three repeated stimulations. *Nereis* does not remain totally unresponsive during a long series of stimulations; the worms react sporadically and it may take as many as 60 or 70 trials before the number of worms that react each time is reduced to 1 or 2 per cent of the population.

Considering the normal tumult in which *Nereis* lives, it is likely that the worm is permanently habituated to a great variety of stimuli. How, then, does it escape its predators? Part of the answer is, of course, that many of the worms do not, but for the rest it seems that they detect the approach of a predator by more complex clues than a simple mechanical disturbance or a shadow passing across them. Habituation to combined stimuli, such as a mechanical shock given at the same time as the light intensity is suddenly decreased, is much slower than to either stimulus alone. Presumably, were one able to reproduce the pattern of stimuli presented to the worm when it is stalked by a crab, habituation would be an extremely slow process. Furthermore, any change in the prevailing pattern of background stimulation is likely to produce reactions. This was shown very clearly by Professor Rullier of the University of Angers, who subjected the serpulid fanworm *Mercierella* to series of 20 stimulations by mechanical shock alternating with 20 by moving shadows. Although the worms quickly became habituated to each, a proportion of them reacted to the first stimulus in a series when the nature of the stimulus changed. The same effect, though less pronounced, can be seen in *Nereis* and if isolated mechanical shocks are interposed in a series of stimulations by moving shadows, the worms are more likely to respond to the shadows immediately after being given a mechanical shock than at other times.

As a variant of habituation to sudden stimuli, S. M. Evans at Bristol has attempted to abolish the reflex behaviour of *Nereis* when it is placed in a glass tube. The tube simulates the burrow of the worm and its normal response is to crawl rapidly to the far end of the tube and search around the opening, but in these experiments the worm was given a slight electric shock or a flash of bright light when it reached the end, either of which caused it to retreat. Learning not to run along the tube involves a higher order of learning than simple habituation, because a time factor (the interval between starting to move along the tube and stimulation) has been introduced. Although worms normally crawl to the end of the tube at approximately the same speed in each trial, those that receive a shock crawl more and more slowly in successive trials, pause frequently, and eventually refuse to crawl along the tube at all.

The biological situation represented in these experiments is the presence of a predator or perhaps some noxious substance at the entrance of the burrow and greater verisimilitude can be got by providing the worm with an alternative pathway, in the form of a T or Y maze, since most nereid burrows have more than one entrance. The worm is shocked if it turns right on reaching the head of the T. Although Fischel claimed some years ago to have trained *Nereis virens* always to choose one branch of a Y maze after only eight trials, Evans has found that the maze-learning ability of several species of *Nereis* is poor. After 20 or 30 trials, the worms may show a statistical preference for the branch of the maze in which they do not receive a shock, but they do not show consistent evidence of learning for very long. An explanation of this may be that the worms have short memories, so that if they make a succession of "correct" choices, the association between "incorrect" choices and punishment begins to wane and the worms revert gradually to random choices until the association is re-established. Another difficulty is that if the worms happen to make a succession of "incorrect" choices at the start of the experiment, they may refuse to run through the maze at all.

In spite of the fact that associations which involve a time element are sometimes difficult to establish and appear to be retained only a short time in *Nereis*, it is possible to establish a conditioned reflex in the worms. In these experiments, worms living in glass tubes are suddenly illuminated, causing contraction in an inexperienced worm and, at the same time, food is presented at the entrance to the tube. The time taken for the worm to crawl to the end of the tube and collect the food decreases with successive trials and by the tenth it comes to the mouth of the tube when the light is switched on whether the food is present or not. The association between illumination and food is retained for periods of several hours in this case, at least overnight, for once it has been trained, its first performance in a day's trials is no worse than any other.

The general impression given by these experiments is that, contrary to expectations, *Nereis* has only a rudimentary ability to learn. Perhaps we have made an unfortunate selection of a subject and, possibly, other polychaete worms living in the inter-tidal zone of the seashore have greater learning abilities. Certainly, some have larger and more elaborate nervous systems than *Nereis*. Perhaps the experiments, if designed slightly differently, might present learning problems more within the worm's capabilities and closer to its normal experience.

But, as so often happens in this type of investigation, one chance observation suggests that we may underrate *Nereis*. When the worms are kept in glass tubes in the aquarium, they frequently fight over the possession of the tubes and an aggresive worm often displaces another worm from a tube and occupies it itself. A worm already in a tube only exceptionally attacks a worm crawling around outside. I have observed this only three times and on each occasion the aggressor, in the tube, had previously evicted the worm it attacked. In no case did the inoffensive worm attempt to enter the tube, nor did the aggressor attack other worms in the vicinity. One rather startling interpretation of these observations is that the worms are capable of recognising individuals with which they had fought some two or three minutes earlier. Other and less invigorating interpretations can be made, but it is this kind of observation that suggests a deeper investigation into the behaviour of *Nereis* might reveal that this worm, with its rather simple nervous system, is by no means as simple as it appears at first sight.

2

Can rats reason?

W. S. ANTHONY
14 November 1957

These early attempts to answer the question could not come up with a convincing answer either way. But the evidence that rats could make use of the abstract concept "lighter than" suggested that some forms of reasoning might not be beyond them.

It is certain that rats can *learn*.

Suppose you put a rat in a maze every day just before his feeding time. If he repeatedly finds the food in a certain place, he will run the maze more and more efficiently (reaching the food after fewer and fewer wrong turns); he is gradually learning to avoid the blind alleys. But this is not regarded as evidence that he can *reason*. To try to show that he can reason, we shall have to see whether he can solve a problem at the first attempt; it must be new to him, otherwise any solution could be attributed to learning. If we use rats that have been brought up in the laboratory, we can be sure that the problem will be new. Furthermore, laboratory rats are healthy, happy and never (hardly ever) bite.

Figure 1 is a diagram of a rather simple maze. The boxes (A, B, C and D) and the pathways (AB, BC, AC and AD) have walls, so that the rat cannot get out, or see across from one part of the maze to another; there is food at C. Suppose we first give a rat experience of the maze, letting him run to C from the other three boxes and then, on what will be called a "test run", put him at A with a block in the direct path to food, AC. The rat is now faced with a slightly new situation: AC, the preferred (because shorter) path to food, is for the first time blocked. Will he now choose ABC, or AD, the blind alley? Actually, out of 48 rats, 75 per cent chose the correct path (ABC) and 25 per cent the blind alley (AD) – this is significantly different from the 50:50 score which would be expected from chance. But although the situation is new, this positive result would still not be considered evidence for reasoning – in training, the rats acquired a tendency to choose AC most frequently (84 per cent of choices), AB infrequently (12 per cent), and AD least frequently (4 per cent), so the test run merely shows that the block did not disturb the preference for AB over AD that already existed.

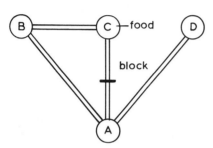

Figure 1 *A simple maze. When the block is in position, the rat must go from A to C via B to get the food*

Suppose that, instead of letting the rat explore the whole maze, we had given him training only on separate parts, allowing him to go from A to B and from B to C but never from A to B to C in one go; his way of getting from A to C will always have been AC, never ABC. Now, the test run, when we block AC, might well be called a test of reasoning ("A to B, B to C, so A to B to C"); what simpler test of reasoning could be imagined? But here John B. Wolfe found that, out of 11 rats, only five chose AB, the other six choosing the blind alley. Other experiments of the same kind have yielded similarly inconclusive results.

Note, however, that the piece of "reasoning" that is required – the "integration of isolated experiences" as it is called – is not all that logical. The rat has been from A to B (unhungry), and he has been from B to C (hungry), but he has never been allowed to go from A to B *and then to* C; in training on AB, BC has always been blocked at the entrance. It might be thought that this would not disturb the rat but there is definite evidence that rats are disturbed by blocks in paths, even if the path is blocked only when the subject is unmotivated for the reward at the other end. That is, if you block the path to food when the rat is not hungry, then he will be less likely to try that path when he *is* hungry. Until a way of avoiding this inhibition has been found, it would be expected to diminish any tendency to choose the correct alley that the rat might otherwise possess.

Another type of problem has been devised, which, as it happens, escapes this objection (Figure 2). The rat is trained to run for food from the start, S, to the goal, G. At first, the entry to path 1 (the shortest path) is closed, and he acquires a preference for path 2 over path 3, path 2 being the shorter of these; next, path 1 is opened, and he now acquires a preference for path 1.

For the test, the entry to path 1 is open, but the door near G is closed. This renders both paths 1 and 2 useless; path 3 is now the only open path to the goal. The rat initially runs up path 1 to the closed door, where he is removed, but after one or more of these abortive runs, he abandons path 1 and chooses either path 2 or path 3. If he chooses path 2, the path which has previously been preferred to path 3, this is a silly response, for it leads him back to the closed door again; but path 3 avoids the door. Actually, out of 29 rats, only six chose path 2 after first abandoning path 1; the other 23 chose path 3. As all the rats had previously shunned path 3, their choice in the test suggests that they somehow understood that path 2 was now useless.

Figure 2 *In this maze, S is the start and G the goal. When path 1 is blocked at A, path 2 is the shorter of the two open paths; when it is blocked at B, path 3 is the only open path*

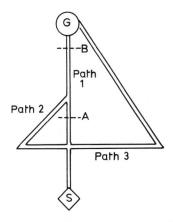

But consider the performance of another group of animals. These rats were trained in the same way, but in the test, path 1 was blocked in front of the point where paths 1 and 2 converge. Path 2, the previously habitual path, was still open, yet nearly half the 26 rats chose path 3. The success of the first group of rats now looks less impressive; indeed, the experimenter's major conclusion was simply that established habits could be quite flexible. Nevertheless, the difference in performance of the two groups is statistically significant, so this suggests that the rats may have, to some extent, grasped the fact that a block behind the point where paths 1 and 2 converge renders path 2 useless, whereas a block in front of this point does not.

This experiment, by Sol Evans, needs further investigation. It is a modification of a problem invented by Edward C. Tolman and C. H. Honzik. Many experiments have been performed on this problem, with varying results, but only Evans's experiment avoids certain objections to the original procedure. But even if rats are found to be capable of solving this problem, it may perhaps be thought that the test is too simple, and does not require the rat to "reason".

I have now summarised what I consider to be the most relevant evidence. One's main impression is one of surprise that so many rats should behave so stupidly in tests whose solutions appear so obvious: 25 per cent of the rats who had had plenty of experience of the whole maze of Figure 1, when the shortest path to food was blocked, ran up the blind alley! Yet, certainly in the ordinary way, rats are perfectly capable of distinguishing between the various alleys of such a maze.

You may believe that one could tell whether rats can reason by simply watching them – do they look puzzled, scratch their heads, appear to make decisions after long thought? The answer is, certainly, but their eventual choice after all this apparent medita-tion is often quite stupid (for example, choosing the same wrong path for the third time in succession). In other words, looking puzzled or enlightened is not evidence of rationality in the rat.

We may sometimes feel that we can never assess the private consciousness of a rat (like that of other groups of which one is not a member – girls or Chinese). This is not a very important difficulty, since the time-honoured test of mental ability is the production of a correct solution and the study of even human consciousness has been comparatively unfruitful. It must be admitted, however, that our common rules for attributing particular mental qualities to humans are vague, and for attributing them to animals, even vaguer. Thus, apart from our ignorance about rats per se, the answer to the question, "Can rats reason?" must be vague for two reasons: (a) the "man in the street" has not decided sufficiently exactly what he means by "reason", (b) the psychologist has not yet acquired enough knowledge to formulate a more precise concept which he could confidently try to persuade the "man in the street" to adopt.

Instead of "Can rats reason?" people sometimes ask, "Can they *abstract*?" The answer to this seems to be yes. A rat can be trained to avoid a very dark alley and to run down a moderately dark alley; if a test is then given, in which he has the choice of the moderately dark alley and a well-lighted alley, he will in general choose the well-lighted alley, avoiding the alley which he had previously preferred. In other words, he responds to the abstract property "lighter than". Although this result is very general, nevertheless rats can also be trained to respond to absolute brightness.

"How far that little candle throws his beams!" Portia observed, coming home at night, and that really contains the seed of any reasonable answer to the question, "Can rats reason?" at the moment. A rat is very dim compared to a man, and cannot do anything that we think of as "reasoning" when we talk of a man reasoning. It may not even be capable of anything that we feel inclined at present to call "a primitive kind of reasoning". On the other hand, it is brighter than would be expected from simple conditioned-reflex theory. "Reasoning" may be a side issue; the experiments referred to here help to show just how the rat does behave (his flexibilities and inhibitions), and may possibly contribute to the forming of more accurate ideas with which to describe "seeking" behaviour in general.

3

A short lesson in animal logic

'MONITOR'

5 June 1975

Psychologists schooled in conditioning theory believe that associations between events are strengthened or weakened according to how often they occur together and for how long. This experiment showed that rats might see things differently.

A quiet revolution is under way in the study of Pavlovian conditioning. Hitherto, most research has attempted to specify the nature of the associations between events that develop during simple conditioning. By contrast, little or no effort has been directed to identifying the internal representations of events which then enter into associations. This was a serious omission since both tasks are prerequisites to understanding how simple learning mechanisms enable animals to construct models of external reality.

A recent paper from Yale University provides a striking illustration of the new Pavlovian research. Robert Rescorla and Donald Heth, writing in a new section of the *Journal of Experimental Psychology*, 'Animal Behaviour Processes' (vol. 104, p. 88), describe recent fear conditioning experiments with rats. They paired a neutral stimulus such as a tone with a mild electric shock for a few trials and then presented it alone for several more. As expected, the tone initially elicited fear, which subsequently declined. The traditional human interpretation of this finding is that the rats develop some association between the tone and shock during conditioning which then weakens when the tone is presented alone.

However, as the two workers point out, rats may not see it this way. Clearly these animals have no sophisticated understanding of electricity. They may then hold to the alternative, and equally valid, notion that shocks are events that come after tones and that they initially hurt but later they do not. Experimental manipulation of the relationship between events may simply change an animal's perception of the events themselves.

To test this interpretation, Rescorla and Heth gave the rats a single "reminder shock" on its own at the end of the experiment when the tone no longer elicited fear. This resulted in a full conditional response when the tone was presented subsequently. The test shows that the association between tone and shock may not have been weakened when the tone was presented alone during the second phase of the basic experiment. Other experiments back up this interpretation and show that the effect is not due to rapid associative learning or nonspecific sensitisation.

Like a lot of good research, these findings raise more questions than they solve. While this work implies that animals form discrete representations or memories of events like electric shock which can change very rapidly – perhaps in one reminder trial – the reason associative connections change more slowly is unclear. Moreover, the boundary conditions governing memory change are as yet uncharted. One of Rescorla and Heth's incidental findings was that their animals' memory of shock could be altered by brief presentations of a loud noise. The boundaries encircling substitutable events may be so broad that one can imagine experiments in future where, through careful training, rats might be taught not only that shocks do not hurt but that, for example, they taste nice.

4

In search of a new theory of conditioning

NICHOLAS MACKINTOSH

26 January 1984

The increasing refinement of experiments involving conditioning has revealed that more goes on inside the animal's head than straightforward associations between stimuli and responses. Not surprisingly, the sort of learning of which a rat is capable is rather better suited to life in the wild than to the confines of the laboratory. But we still do not know enough to formulate a new theory of this type of learning.

The Russian physiologist Ivan Pavlov first discovered and made a serious scientific analysis of the phenomenon of conditioning, having already been awarded the Nobel prize for his research on the digestive system of dogs. Pavlov found that a restrained, hungry dog would start salivating not only when dry food was placed in his mouth or at the sight of the food but also at the sight or sound of the attendant who usually provided the food and, eventually, in Pavlov's experiments during the first two decades of this century, to any arbitrary stimulus such as a flashing light regularly preceding the delivery of the food. It is something of an exaggeration to say that no one before Pavlov had ever observed conditioning but it is certainly no exaggeration to say that he both provided the first sustained scientific analysis of the phenomenon of conditioning and developed the terminology with which this phenomenon is encumbered to this day. Salivation to food, in that terminology, is an *unconditioned reflex* but the salivation then elicited by the flashing light is a *conditioned reflex*, dependent on the animal's particular experience, being strengthened or *reinforced* by the presentation of food immediately after the illumination of the light and weakened or *extinguished* if subsequent appearances of the light are not followed by food.

Much of this terminology, indeed, was subsequently applied to the results of another set of experiments, undertaken at much the same time as Pavlov's but in total ignorance of his work, by the American psychologist Edward Thorndike. Thorndike was more eclectic in his choice of animal than Pavlov, using chickens, cats and monkeys impartially and the impetus for his work was also different. Whereas Pavlov regarded himself as a physiologist studying the brain, whose task it was to infer the principles governing the organisation of higher cortical structures from observation of the behaviour they controlled, Thorndike was concerned first to provide objective evidence about the ability of animals to learn and solve problems. He wanted to counteract the widespread reliance on anecdotal evidence of the kind amassed by George Romanes in his book, *Animal Intelligence* (published in 1882) but above all to resist the tendency to provide anthropomorphic interpretations of that behaviour – a tendency that owed much to Romanes's desire to prove the mental continuity of man and other animals. Thorndike argued that, when they were studied objectively, evidence for anything approximating to human intelligence in animals evaporated and that even monkeys solved the problems he set them by a process of blind trial and error rather than by reasoning or insight. He placed animals in a puzzle box from which they could escape and gain access to food by operating a hidden catch, pressing a latch or pulling on a piece of string. Chickens, cats and monkeys, Thorndike claimed, all solved this sort of problem rather slowly and at similar rates. He interpreted their behaviour in terms of his celebrated "law of effect", formulated in 1898: the reward of escaping from confinement and obtaining food served to strengthen the one response that was successful in opening the door, while all other responses, being followed by no desirable effect, were weakened.

Thorndike's work was for a long time, at least in the West, rather more influential than Pavlov's. In the hands of the Harvard psychologist, B. F. Skinner, the puzzle box was turned into a Skinner box, the main difference being that the food reward was now delivered to a recessed tray or magazine in one wall of the box whenever the animal performed the required response and it was possible, therefore, to confine animals to the box for several hours at a time. Skinner's favoured animals were rats, required to press a lever, or pigeons, required to peck at a small illuminated disc, and these animals have remained the most frequently studied subjects in experiments on instrumental learning or *operant conditioning* as Skinner called it. Skinner used Pavlov's terminology of condition-

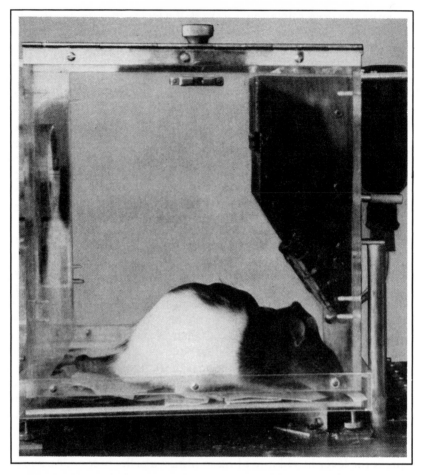

The laboratory rat: probably the most studied, but not the best understood, member of the animal kingdom

ing, reinforcement and extinction but deliberately eschewed physiological theorizing. His central idea was borrowed directly from Thorndike's law of effect: responses were strengthened or reinforced if they were followed by one kind of consequence (a positive reinforcer) and weakened or extinguished when no consequence followed.

In the hands of Clark Hull, from Yale, Thorndike's theory of conditioning came to dominate American psychology for decades, to an extent that is hard to credit today. Research on conditioning and learning (the two terms being virtually interchangeable), whether on dogs salivating, rats pressing levers and running through mazes, or college students learning lists of nonsense syllables, was all conducted within the same framework. Learning or conditioning was reducible to the automatic strengthening or weakening of stimulus–response (S–R) connections: knowledge, purpose and foresight were concepts that could be safely dispensed with and the principles that worked so successfully in these simple and contrived experimental situations would equally explain how young children learned their native language or older children differential calculus.

Few psychologists would subscribe to this dream today, if only because it has become apparent in the past 25 years that Thorndike's theory is quite inadequate to explain even the results of simple conditioning experiments. Conditioning, it now seems, is best viewed not as a matter of strengthening S–R associations by an automatic process called reinforcement, but as one of detecting contingencies in one's environment, learning about relations between events, or finding out what causes or signals the occurrence of events of consequence such as food or water, danger or safety. The response of salivating or pressing a lever is not what is learned in a conditioning experiment; it simply provides a convenient index of what the animal has learned – that the light signals the impending delivery of food or that pressing the lever will produce a pellet of food. What is the evidence that has led to this change of view?

There was one line of experiment that had always created problems for Thorndike's and Hull's analysis. More recent variations on this theme have forced a rejection of their account. Suppose that a hungry dog has been conditioned to salivate to a light which signals the delivery of food, or that a rat has been conditioned to press the lever in a Skinner box to obtain a pellet of food. According to Thorndike and Hull, all that the animals have learned is to perform a particular response whenever a particular stimulus is turned on. What would happen, then, if the animal was made to dislike the particular food previously used to reinforce conditioning? There is a simple and effective technique for such *conditioned aversion*: animals eat the food and shortly thereafter receive an injection of lithium chloride. The injection makes the animals ill and, as Christopher Adams and Anthony Dickinson at Cambridge have shown, they rapidly learn to avoid eating this particular food. They attribute their illness to the food they have eaten (its most probable cause in the natural order of events), rather than to the injection.

The question is whether, put back in the conditioning stand or

Skinner box, the dog or rat will continue to salivate or press the lever whenever the light is turned on. Common sense says no; the rats, for example, should no longer press the lever when the sole consequence of doing so in the past has been to produce a pellet of food which they would now refuse to eat because they have learned it makes them ill. But S–R theory says otherwise. According to Thorndike, the rats do not press the lever because they have learned that lever pressing produces food. They are not credited with knowledge about the consequences of their actions. Similarly, the dogs do not salivate because they believe that the light signals food; they have just been conditioned to salivate whenever the light is turned on. The light elicits the response of salivating or lever pressing and will continue to do so until some new, adverse consequence has followed the sequence of light followed by response sufficiently often to outweigh this old habit. Pairing the food alone with some new consequences will have no effect.

By and large, the results of such experiments have been a victory for common sense. The rats refrain from pressing the lever and it must be supposed, therefore, that they learned, first, that lever pressing had a particular consequence, secondly, that this consequence was no longer desirable, and finally were able to put these two pieces of information together. It is worth stressing that they do not always behave so sensibly. There are circumstances under which rats will continue to perform a response whose previous consequences they no longer value. The response, one is inclined to say, has become a habit, triggered off whenever the animal is put in a particular situation, without regard for its consequences. The description captures the point of S–R theory: it is a theory about habits, or behaviour that is run off whenever the appropriate stimulus appears, unguided by any thought of its consequences. Just as we know that habits are formed by constant repetition of the same action in the same circumstances day after day, so it is that the rat which has received extensive and unvarying training to press the lever for the food pellet is the most likely to continue doing so even after an aversion has been conditioned to that food.

One tenet of traditional theories of conditioning, accepted by Pavlov as much as by Thorndike and Hull, was that successful conditioning depends critically on close temporal contiguity between stimulus, response and reinforcement. Indeed, the role of temporal contiguity has been stressed by the classical associationist philosophers, by Aristotle, David Hartley and David Hume, as much as by 20th-century theories of conditioning.

Animals in conditioning experiments, however, pay little attention to traditional theory. Close temporal contiguity may be one important variable in conditioning but it is neither necessary nor sufficient. That it is not necessary is most forcefully demonstrated in experiments on the conditioning of food aversions. As we have seen, a rat given an injection of lithium chloride will associate the illness induced by the injection with any food or drink recently consumed and thus form an aversion to it. John Garcia, then at the State University of New York at Stony Brook, demonstrated a most striking feature of this conditioning and one which makes the use of slow-acting rat poisons a waste of time and effort: the interval elapsing between consumption of food and illness can be several hours without abolishing the aversion. The length of the interval has some effect: other things being equal, rats made ill within 30 minutes of drinking a novel-tasting substance will show a more marked aversion to that substance than those made ill 300 minutes later. But even at this longer interval some associative learning has occurred. The rat shows an aversion to the food and appropriate control experiments suggest that the aversion was conditioned, rather than, for example, being an example of neophobia (a tendency to avoid any novel substance) made stronger by the experience of illness.

Food and sickness are not the only events that do not require strict contiguity for an animal to associate them. Under certain circumstances rats can learn which of two alternative pathways to choose in a maze, even though an interval of several minutes elapses between their choice on each trial and the outcome of that trial. For example, Bow Tong Lett of Memorial University, Newfoundland, removed rats from the maze immediately after they had chosen on each trial, regardless of whether their choice was right or wrong, and only later returned them to receive food in the maze if their choice was correct. They learned successfully with delays of at least 5 minutes between correct choice and reinforcement.

The association span of the rat, it appears, is capable of bridging quite long intervals. But this finding immediately poses a new problem: why does it not always do so? The belief that successful conditioning requires close temporal contiguity between the events to be associated was not, after all, a complete myth. Numerous careful experiments had looked for evidence of conditioning across intervals of more than a few seconds without success, and their results made functional sense. If we assume that the function of conditioning is to enable organisms to find out what signals or produces certain events of consequence, in other words to attribute the occurrence of, say, food or danger to their most probable antecedent causes, it makes no sense for an animal to associate the occurrence of food with any and every event that has happened in the preceding hour. Rather few of these events will in fact be causally related to the delivery of food and the animal's task is to distinguish between probable causal relationships and occasional chance conjunctions of events. The mechanisms of conditioning, it turns out, are nicely designed to achieve just this.

The critical experimental observation is this. It is true that successful conditioning can occur to a stimulus even though there is a significant temporal interval between the occurrence of that stimulus and of the reinforcer, or even if the stimulus is only imperfectly correlated with the delivery of the reinforcer (as when the reinforcer occurs on only a randomly chosen 50 per cent of trials on which the stimulus appears). But, and this is the critical point, such conditioning will be prevented if there is some other event that predicts the occurrence of the reinforcer more successfully. Conditioning occurs selectively, to good predictors of reinforcement at the expense of poorer predictors. For example, Ben Williams of the University of California at San Diego showed that pigeons would learn to peck at a disc briefly illuminated with red light even though he imposed a delay of 10 seconds between such pecks and the delivery of food. But such learning was effectively abolished if some other stimulus, for example a brief green light, occurred in the interval between a peck on the red disc and the delivery of food. It is as if the pigeons attribute the food to the more recent occurrence of the green light rather than to their earlier peck on the red disc (Figure 1). Rats will learn an aversion to saccharin if they are made ill (by an injection of lithium) shortly after drinking it – even though they actually receive the injection on only 50 per cent of occasions on which they drink the saccharin solution. Maintaining exactly the same, imperfect, relation between sugar solution and illness, Alexander Luongo of the University of California at Los Angeles arranged that the rats should also drink a saline solution on days when lithium was to be injected. The rats attributed their illness to the saline rather than to the saccharin solution, and showed no aversion to the latter.

So whether or not a particular event is associated with a particular reinforcer does not depend solely on the relationship holding between the two, as traditional associative theories have implied; it also depends on whether *other* events are in a more favourable

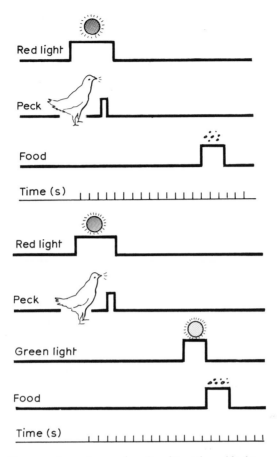

Figure 1 *Pigeons learn to peck a disc lit with red light even if they have to wait 10 seconds for food. But they give up pecking if a green light appears in the interval, attributing the food to the green light rather than to their pecks*

position to be associated with that reinforcer. If there is no other event more reliably related to the occurrence of the reinforcer, then the poor predictor remains the most probable cause and is successfully associated with the reinforcer. But if there is a better predictor available, then the less reliable relationship is relegated to the status of coincidence and produces no conditioning. Thus, the reason why even moderate intervals between stimulus or response and reinforcer are often sufficient to prevent conditioning is that it is likely that some other event will occur during the interval and will itself be associated with the reinforcer at the expense of the earlier stimulus or response. Successful conditioning over long delays requires that no such potential interfering association be allowed to occur. This is relatively easy to ensure in the case of food aversion since rats are selectively biased towards associating illness only with something they have eaten or drunk. By denying the rat access to other food or drink during the interval between exposure to sugar solution and an injection of lithium, the experimenter can expect to see conditioning of an aversion to the sugar solution across intervals of several hours.

One final feature of the conditioning process also serves to ensure that conditioning occurs more readily to events that are likely to be causally related to the occurrence of reinforcers. An animal's past experience of the significance of a stimulus can profoundly affect the extent to which the stimulus will be associated with reinforcement in the future. Conditioning will occur only slowly to a stimulus that has occurred repeatedly in the past without ever being followed by any other event of consequence. Thus rats made ill by an injection of lithium will not normally condition an aversion to the food and water that have been their daily diet since they were weaned. If two events have tended to occur independently (uncorrelated with one another) in the past, animals will subsequently associate them only with great difficulty. An animal whose efforts to avoid unpleasant events have hitherto been to no avail may be extraordinarily slow to learn that its actions now do have positive consequences (a phenomenon sometimes termed "learned helplessness"). Or, if an experimenter arranges that a pigeon placed in a Skinner box receives food from time to time and, uncorrelated with the delivery of the food, a green light occasionally appears, the pigeon will be very slow to associate the green light with the delivery of food when the experimenter subsquently establishes a positive correlation between the two. The conditioning process is designed on the assumption that the world is a moderately stable place: other things being equal, events that have been unrelated in the past are unlikely to be causally related today.

It is one thing to throw out an inadequate theory; it is another to replace it with a better one. We know enough about the nature of conditioning to say that it cannot readily be accounted for in terms of traditional S–R theories; we do not know enough to be able to specify the details of a new theory. There is no shortage of ideas. One key notion is that conditioning depends on the predictability of the reinforcer: if the occurrence of the reinforcer on a given conditioning trial is already predicted by one event, little or no conditioning will occur to other events present on that trial. But this is an empirical generalisation open to more than one theoretical interpretation. One possibility is that the concept of reinforcement should be reinterpreted to signify the discrepancy between obtained and expected reinforcement. Another is that organisms learn to ignore redundant events that signal no change in reinforcement. But this degree of theoretical uncertainty is surely a sign of life. The study of simple associative learning or conditioning in animals has progressed a long way since the time of Pavlov and Thorndike. Conditioning may be relatively simple by comparison with other forms of intelligent behaviour. But it is not as simple as earlier theorists would have had us suppose. The image of Pavlov's dog helplessly drooling at the sound of the bell, or of a rat trundling its way through a maze by a process of blind trial and error, is singularly inappropriate. The process of conditioning is one which enables animals to establish a subtle and accurate picture of the causal sequence of events in their environment, to predict where and when food or danger can be expected and to modify their behaviour accordingly. The processes by which they achieve this remain imperfectly understood, but a challenge to future experimentalists and theoreticians.

PART TWO

Animals in the Field:
The Art of Observation

By contrast with the rigid and regimented activities of the behaviourists, the ethologists came as a breath of fresh air. Their prewar observations of animals in the wild were of necessity less controlled, and allowed a certain amount of licence in the matter of interpretation. Konrad Lorenz's delightful anecdotes in the book *King Solomon's Ring* are today rather frowned upon as a means of communicating scientific information. But he and his fellow Nobel prizewinners, Karl von Frisch and Niko Tinbergen, demonstrated the importance of watching what animals actually do as they go about their daily lives. Only when you really know *what* they do can you make tentative hypotheses about *why* they do it and think of ways of interfering with the behaviour in order to test your hypotheses.

In the early days the ethologists looked for causative answers to the question "why", that is, answers in terms of some internal driving force. More recently they have become virtually obsessed with function – evolution's motives, rather than the individual's. Either way, they need data in the form of many hours of observation in order to say anything sensible. Conventional beliefs about animals might turn out to be quite mistaken when they are studied systematically. The hyaena is a case in point. In one natural history book after another hyaenas were described as scavengers, too cowardly to kill any but the very young or the very sick. After years of patient watching, Hans Kruuk was able to present the hyaena in a different light – as a skilful hunter with a sophisticated social structure that allowed close co-operation between individuals in making a kill (p. 35).

Kruuk's research faced all the difficulties of keeping track of a wild population spread over hundreds of square kilometres. Those interested in the behaviour of rhesus monkeys have been luckier. These monkeys are natives of India, where they now tend to live in close association with human settlements. But since 1939, a colony of rhesus has lived a comparatively undisturbed existence on the island of Cayo Santiago in the Caribbean (p. 4) – undisturbed, that is, except by waves of primate researchers who have come to recognise almost every individual and have witnessed the power struggles that take place among the monkey troops.

Another long-term study that is seeking to answer crucial questions about evolution and behaviour is the Rhum red deer project, undertaken by a team from Cambridge University (p. 50). Again it is crucially important to recognise individuals in order to find out what are the characteristics of the most successful breeders. Similar considerations apply to the well-known great tits of Wytham Wood near Oxford, which have been under observation for more than 30 years (p. 61).

Although the ethologist's best assets are a good pair of binoculars and a notebook, a bit of technical wizardry does no harm. On the contrary, some observations would probably never have been made without the assistance of technology. David Lack took advantage of the echoes made by birds on radar screens to come to the conclusion that swifts took off on summer evenings and might spend the whole night on the wing (p. 67). Modern techniques of radio tracking and biotelemetry mean that it is no longer necessary to be within sight of an animal to know where it is or even what it is doing, a considerable advantage when dealing with shy or nocturnal species (p. 75). These are just two examples; other invaluable techniques include sound spectrography, which enables the sounds of animals and birds to be represented visually, and high-speed photography to observe fast-moving species at a more leisurely pace.

5

Hyaenas as hunters

HANS KRUUK

30 June 1966

Hyaenas were long thought to be the refuse collectors of the African savannah, living on the leavings of nobler beasts. But the Serengeti research project revealed them to be skilful hunters in their own right, with an important place as predators in the ecology of the region.

For a variety of reasons, it might in the future become necessary to interfere with the large populations of game in the Serengeti plains of Tanzania. Fortunately, we cannot now envisage a direct need for this, but we may have to face this necessity – for instance, to meet growing local demand for protein, or as management measures in order to stabilise population trends. Such interference with the course of nature would be dangerous if not preceded by careful research into the population dynamics of the species involved, as well as into the relations of these species with their habitat and study of the natural sources of mortality. For this reason, the Director of Tanzania National Parks initiated the Serengeti Research Project.

The study of predators acting on the plains' game forms an important part of the programme of the Serengeti Research Institute. Lion, leopard, cheetah and wild dog are all fairly common in the area, but by far the most numerous predator is the spotted hyaena (*Crocuta crocuta* Erxleben). This species was therefore singled out as the special subject of the present study.

The role of hyaenas was for long believed to be that of scavenging the remains of game killed by other causes, but several incidental observations had already given rise to the suspicion that they might also attack and destroy large numbers of very young animals. Yet even if they were only scavengers, their role in the ecology of the Serengeti would be an important one.

The research focuses on the Serengeti National Park and its direct surroundings, but early on I found it was necessary to study hyaenas elsewhere as well. In the nearby Ngorongoro Crater some hyaena activities could be observed much more easily; the 100-square mile floor of the Crater is sealed off almost entirely from the surrounding country by the high walls, and the large resident population of hyaenas is far less shy than in the Serengeti.

A family group of hyaenas in the Ngorongoro Crater

In 1964, during the first months of our stay in East Africa, my observations tended to confirm the conventional ideas on the feeding behaviour of hyaenas. Whenever I saw hyaenas hunting or eating they were usually concerned with young Thomson's gazelle, and their clear ability to spot and to react to alighting vultures indicated an adaptation to a scavenging way of life. But, gradually, more and more information indicated that this was by no means the entire story. Only after I learned how to follow the hyaenas on moonlit nights could I build up a complete picture of their hunting behaviour, which was quite different from what was generally thought about them.

During daytime, hyaenas sleep in holes or crevasses, in the mud somewhere, in the shade of a tree or sometimes just out on the open plain. At dusk, they leave these places, walking slowly round, meeting other hyaenas. They spend a seemingly interminable time in greeting ceremonies and gradually a concentration of hyaenas builds up (a handful, perhaps, or as many as 20) usually somewhere near one of their communal den-sites.

Thus, the hyaena "pack" is formed and when they all walk off together, they may keep close or deploy far apart, often not showing the slightest interest in the many animals around them. They go in a fixed direction, often playing or depositing scent marks on the grass. Then, sometimes after walking miles in this fashion, their behaviour will change when they spot their quarry. If it is zebra, the hyaenas are after that night, they will slowly walk right up to the group of zebra, heads rather low, tails slightly raised, sometimes as near as four or five metres.

Zebra live in families, consisting of one stallion, up to six or seven mares and some immatures. When the hyaenas approach, the whole family stands, watching. The stallion usually attacks when the hyaenas are near, with head low and teeth bared, snapping at the hyaenas and lashing out with his legs where he can, while the rest of his family flees. In the ensuing chase, the typical picture is that of a dense group of zebra mares and young running fairly slowly (sometimes even stopping), followed by the rather sparse hyaena pack in a kind of crescent formation. The stallion dashes backwards and forwards between the family and its pursuers.

Every hyaena seems to work regardless of the other members of the pack, and several try to bite at the legs or the soft parts of a zebra. Suddenly one of the zebra (often an adult in the prime of life) may fall back, being hampered in its movements by one or more biting hyaenas. Immediately all other hyaenas concentrate on that animal, biting at the loins and anal region, and, whilst doing so, preventing the victim for moving any farther. The zebra then stands without defending itself, falls over after a few minutes and dies usually within 10 minutes after being caught. Half an hour later, there is not even a bone to mark the spot.

The hyaenas are not always successful, by any means. Of the 33 occasions when I was able to watch the beginning and end of a zebra chase, only six times did the hyaenas make a kill. But on several other occasions, I have come on to the scene when a zebra has already been caught. We do not yet know what determines the hyaena's success. It may well be that zebra families that run away faster than others are more likely to escape (it seems that the zebra's low running speed when attacked by hyaenas is related to the fact that the stallion does not lead the family but follows).

The hunt of wildebeest is a slightly different affair for the hyaenas, because the wildebeest do not live in families like the zebra but are either territorial males or live in large herds. Hyaenas approach a herd of wildebeest in the same way as they do zebra. When the wildebeest wheel round and run, the hyaenas charge right in between them, each going its own way. Once one hyaena has a firm grip on a wildebeest, the others seem to detect this immediately and converge on the victim, killing it as they kill a zebra. The wildebeest runs much faster than the zebra when it is chased and once caught it is far more likely to put up a fight and charge with its horns. A peculiarity is that wildebeest, while being chased, often run into the water of a brook or a lake, where they are caught almost immediately. On the 26 occasions that I came across hyaenas on a wildebeest kill in the Ngorongoro, in eight cases the victim had been killed in the water.

More than 1100 times I have watched a hyaena eat in the Ngorongoro Crater. In 57 per cent of the cases this was on adult wildebeest, and in 28 per cent on zebra. This is not very different from the ratio in which these species occur there – approximately 17 000 wildebeest and 6000 zebra.

We might be tempted to conclude that the hyaena predation pressure on the two species is about equal. That would be jumping to conclusions. In the period that I saw 18 wildebeest hunts, I came across 33 zebra hunts and during most of the nights that I stayed with the packs it was clear that they paid hardly any attention at all to the large herds of wildebeest around, but would walk for many miles to get near the zebra. It seems, therefore, that the predation pressure by the hyaena population is much more strongly directed towards zebra than towards wildebeest, but apparently the zebra are better able to cope with the situation. The Serengeti data largely confirm these ideas, but are more difficult to interpret because of the migratory habits of the game on the plains.

Many hundreds of hyaena faeces have been analysed and it was, for instance, by this method that the importance of scavenging was demonstrated. One group of hyaenas has its den about two miles from a Masai village in the Crater and I found that, in the faeces of those hyaenas, hair of cattle or goat or sheep occurred in 55 per cent, of wildebeest in 59 per cent and of zebra in 21 per cent. The hyaenas in other parts of the Crater, at the same time, had hair of domestic stock in only 0.5 per cent of their faeces, of wildebeest in 83 per cent and of zebra in 46 per cent. This difference is certainly not caused by a lesser abundance of wildebeest and zebra in the hunting range of the hyaenas first mentioned; it must then indicate a great preference for scavenging around a human settlement. It surely is this habit, together with the fact that hyaenas in daytime are more likely to eat other animals' kills or to kill young small animals than they are at night, that has given them their bad name as "mean" beasts.

The data from faecal analysis will be used to establish accurately the diet composition, and from following known groups of hyaenas, I hope to find their average daily consumption. But to assess the hyaena's predation pressure on the game population, we not only have to know characteristics of their diet, but also their density and distribution. Because of the animal's nocturnal habits, ordinary game counts reveal very little about the numbers of hyaenas in an area and I, therefore, adopted a mark-recovery scheme. From the Land-Rover one can dart hyaenas with a syringe of sedative from a dart gun and, in this way, 50 were captured in Ngorongoro and 200 in the Serengeti. After the dose of succinyl-choline took effect, I marked the hyaenas by clipping triangles out of different parts of their ears. These earmarks are extremely effective and allow recognition of different marked individuals a long way off.

In the Serengeti, most of the animals were marked in March of this year, so not many have been sighted again at the time of writing. But in the Ngorongoro, I have been able to follow the marked population for over a year now and the first results are emerging. I counted the proportion of marked animals that we came across in the various parts of the Ngorongoro and plotted the marked individuals on maps. But also, at various times of the day and night, we attracted hyaenas to the Land-Rover by playing to them, with a large loudspeaker, the tape-recorded sounds of hyaenas on a kill. This might bring as many as 60 hyaenas to the car, and from all these observations it was found fairly consistently that slightly more than one in eight of the adult hyaenas in the Crater had been marked. So this settled the total hyaena population there at about 420 adults and, probably, some 90 younger animals.

Comparing consecutive observations of each individually marked hyaena, we notice that they do not move around at random but stay within the same range. The ranges of various hyaenas either overlap completely or hardly at all; we can distinguish clear groups within the population. Those groups of hyaenas living on one range I called "clans" and it turned out that there were eight of these in the

Ngorongoro, varying in numbers from 10 to about 100. The females especially adhere very strictly to the clan range and, out of the 202 occasions that a marked female was sighted again, only three times was she in a different range from the previous observation. But the males are less strict, and of the 147 resightings, 24 were concerned with a male that had the previous time been seen in another range.

The clan ranges are not just a range, they are also territories, defended by the resident clan. It happens occasionally that, after a long chase, a wildebeest or zebra is killed within the range of a neighbouring clan. If the members of that clan notice what is happening, a struggle may ensue, involving sometimes more than 70 hyaenas. Members of the same clan perfectly tolerate each other on the kill and around the den but, as soon as different clans are, involved, there is threatening and fighting. In fact, one sees actual fights only rarely – one of the groups involved usually gives in immediately. But there may be very noisy encounters in which the hyaenas from different clans chase each other over large distances, biting if they can.

The outcome of a "clan struggle" over a carcass seems determined only to a small extent by the number of hyaenas in the two groups; much more important seems the place of the kill in relation to the territorial "boundary" and the hunger state of the participants.

Remarkably, this system of clan territories existing in the Ngorongoro Crater seems to have broken up in the Serengeti. The

A hyaena threatens another from a rival group

main prey species of the hyaenas migrate through this vast area, and at one time of the year wildebeest and zebra may be over 100 miles away from the place where they were some months before. Only few places have a suitable prey population the whole year round and there we find resident hyaena clans, as in the Crater. There are also clans which have a permanent clan territory but from there make excursions of several days to where the game is, often more than 50 miles away (we call them "commuters"). But most of the hyaenas are simply migratory, following the large herds whenever they go. From the first resightings of marked animals in the Serengeti, I am beginning to get some idea of the complexity of the territorial system there and it is only with the help of observations in the Ngorongoro Crater that I shall be able to interpret my data.

The relation between the numbers of hyaenas and the numbers of their prey species is a very complex one and, because of the hyaena's very common occurrence, it is vital for our understanding of control of animal numbers in the Serengeti to know much more about it. We hope that at least some insight may be gained by systematic study not only of actual numbers, resightings of marked animals and diet composition, but also of the behaviour of hyaenas towards their prey and each other and the behaviour of the prey species towards hyaenas. Only a combination of ecological and behavioural approaches will provide the answers we need.

6

Forty years of rhesus research

RICHARD RAWLINS

12 April 1979

A colony of rhesus monkeys on the island of Cayo Santiago in the Caribbean has given scientists the opportunity to study relationships within a stable monkey community over many years.

The rhesus monkey is, perhaps, the best known of all primates. It was fundamental in the discovery of blood types and was the first primate to be rocketed into the stratosphere. We have detailed information on its anatomy and physiology and a very good idea of its behaviour, and we have had this information for many years. But little of this wealth of knowledge comes from the forests of south-eastern Asia where the rhesus normally lives. Instead, because the rhesus is hardy and easy to breed, we have studied it in the laboratory, rather than in the field.

The island of Cayo Santiago provides a laboratory in the field, where we have gained a great deal of understanding of the organisation of rhesus society and the way the society develops from the growth and evolution of the animals that make it up. The colony on Cayo Santiago is now 40 years old, and this seems a good time to look back on the birth and the development, not of the monkeys, but of the oldest continuously maintained primate colony in the world.

Cayo Santiago is a 15.5 hectare island that lies about one kilometre off the coast of Puerto Rico in the Caribbean. It was founded in the late 1930s at the time when, through the work of men like Robert Yerkes, Solly Zuckerman and Otto Kohler, modern primatology was emerging as a discipline. C. R. Carpenter was a student of Yerkes and he was one of the first behaviourists to go out and watch primates in the wild, both in Panama and Asia. Carpenter, frustrated by the difficulties of field work and the growing demand for monkeys as lab animals, decided to set up a "wild" population that would supply animals for biomedical work and allow long-term observations of behaviour. His initial plan was for a mixed group of rhesus monkeys and gibbons.

At the time, the Indian government was in the habit of periodically slapping an embargo on the export of monkeys. The ones that were freighted suffered terribly, and many died from rampant tuberculosis or diarrhoea. Carpenter got together with G. W. Bachman, Director of the Columbia University School of Tropical Medicine in San Juan, Puerto Rico, and E. Engle and P. Smith of the Columbia University College of Physicians and Surgeons in New York. They obtained a grant of $60 000 from the Mary and John Markel Foundation to begin a research and breeding programme on Cayo Santiago, which at the time was used as grazing for goats. Cayo was to be the first of a series of facilities that would ensure a controlled and regular supply of monkeys for institutions on the mainland.

Bachman hired M. I. Tomlin, a primatologist from the Philadelphia Zoological Park, to manage Cayo, and they set about installing boat docks, watering systems and trails across the island, and the Civilian Conservation Corps began the slow process of reforestation. Carpenter, meanwhile, went back to Indo-China, Thailand and Malaya in July 1938 to collect the Lar gibbon (*Hylobates lar*) and by September he was in India trapping rhesus monkeys in the mountains near Lucknow, about 400 miles from Calcutta. He was quite successful; 450 rhesus and 14 gibbons were released on Cayo Santiago between December 1938 and January 1939, after having first been screened for tuberculosis and marked with an identifying tattoo.

The rhesus organised themselves in the 18 months after they were introduced to the island. There was a lot of fighting and many monkeys died, but when the dust settled there were six social groups. The gibbons were not so lucky. They competed with the rhesus for food and living space, and attacked monkeys and people; they were recaptured, caged and eventually sold. Once the gibbons were out of the way, the work on rhesus took priority and no more animals were added.

Carpenter and Tomlin watched the population into the 1940s, keeping records on who made up the groups, how they behaved socially and sexually, what kind of dominance hierarchy existed and how the troop moved around the island. At the same time, scientists from the School of Tropical Medicine studied sexual behaviour and menstrual cycling, intestinal parasites, and

The rhesus monkey, Macaca mulatta

haematology. The Second World War made it difficult to get supplies out to the island and 490 animals were shipped to the mainland for research on disease. By 1944 there were only some 200 animals left and studies of behaviour had come to an end. The colony was poorly managed and food supplies were irregular in the extreme. The monkeys took to eating what vegetation there was, and the colony and island went into general decline.

The years after 1944 were lean ones for the monkeys on Cayo Santiago but, somehow, the colony survived. Then the island was designated the behavioural branch of the Laboratory of Perinatal Physiology, sponsored by the National Institute of Neurological Disease and Blindness. The animals were used in the lab to study fetal medicine and physiology and were the foundation of many an advance. For students of behaviour, however, a crucial event was the arrival, in 1956, of Stuart Altmann. He reintroduced the regular census of the population, marked all the individuals and worked on the monkeys for two years, eventually producing a masterly study of rhesus behaviour and social organisation that set the stage for all the subsequent studies. John Kaufmann, Carl Koford, M. Varley and Donald Sade continued Altmann's pioneering work and through their efforts the records we have of the Cayo monkeys are probably the most complete for any monkey troops anywhere.

When the Perinatal Lab was disbanded in 1970, Cayo Santiago entered another phase of research as part of the Caribbean Primate Research Center, a new facility sponsored by the National Institutes of Health and the University of Puerto Rico's School of Medicine. Before, the primary purpose of Cayo was to supply monkeys for medical research and behavioural studies were not permitted to interfere with this function. Individuals were selected for "export" haphazardly, with no regard to their social group, age or sex. The result was that many of the groups were extremely unbalanced and artificial, with compositions that were nothing like those in the undisturbed state. One could not even guess what the effects of these unthinking removals were on the patterns of behaviour. Donald Sade was appointed scientist in charge in 1970 and he saw to it that the policies were changed. All but the intact social groups were taken off the island, and the four remaining troops were not disturbed again. Sade stopped sporadic capture and manipulation in an attempt to set up the naturalistic conditions that would be best for long-term studies. Sade's foresight, and the support of the Animal Resources Branch of NIH, allowed Cayo to emerge as a unique site for developmental studies of free-ranging rhesus monkeys, a pre-eminence it retains today.

The island found favour with researchers for many reasons. It is easy to get around and the animals are generally clearly visible. We have accurate estimates of the population age structure and detailed long-term histories of the social and physical development of all the animals. All this is in distinct contrast to the rigours of the field, where unknown animals are invisible most of the time. Nor do we have political terrorists on Cayo! Of course, there are drawbacks too; we give the monkeys provisions, so there are some aspects of their feeding ecology that we cannot study. But I think that the caveats introduced by providing food are more than offset by the special opportunities to follow the behavioural development and population dynamics of a species over many generations.

When Altmann arrived in 1956 there were 150 animals, in two groups, alive on the island. As I write there are 610, all descendants of the stock released 40 years ago, and the once overgrazed island is now heavily forested. With the exception of six solitary males, who seem to dislike company, the animals are in six social groups, ranging in number from 53 to 139 animals. We know the age, sex and maternal genealogy (neither we nor the monkeys can be certain

Young monkeys on Cayo Santiago enjoy swimming and leaping into the sea

of the paternal side of things) of every individual, and have a detailed biography as well. Each year we trap the monkeys once – to tatoo the youngsters born that year and draw blood samples for the continuing genetic work – but aside from this we permit no other interventions.

We especially encourage people to study the rhesus over a long time period because the monkeys have a distinct annual cycle. The way they divide their time between different activities depends on the time of year; broadly speaking we can divide the year into the mating season, July to December, and the birth season, January to June.

Daily activity, too, shows a clear rhythm. The animals look for food and eat in the cool of first light and spend the heat of the day resting and grooming one another on the ground or in the trees. As the afternoon wears on they begin to feed and play again, finally going back up into the trees at sunset to sleep. Much of the day is spent in grooming; this is clearly sanitary in that it helps to eliminate crud and parasites from the skin and fur, but it is also socially vital because it establishes and cements the bonds between members of the group. Indeed, grooming tells us an enormous amount about the social structure of the group; such things as who grooms him, who he grooms, how often, and for how long, provide excellent measures of each monkey's social status relative to the others.

Each social group is made up of a number of adult males and between two and four matrilines. Each matriline, in turn, consists of an adult female, her adult daughters and all their juvenile offspring. Juvenile males leave the troop they were born into when they become mature (about 3–4 years) but the females remain in the natal troop, so that the maternal genealogies form a stable nucleus around which the day's activity is centred. The adult males in a troop, unlike the females, were born into another troop, and will occasionally move from one troop to another, though they never return to their natal troop. So, as in many vertebrates, it is the males who, by their movements, stir the genetic pot, but we do not know what controls the movements of the males.

Within each troop there is social order, with separate hierarchies for males and females (and their offspring). The hierarchy controls access to desirable objects, be they food or grooming partners, and limits the amount of aggression in the group. Rhesus, like all primates and a goodly number of other species, have a well-developed code system for signalling dominance and threat, defeat and subordinance, so there is not too much overt fighting, and aggressive encounters are rarely in themselves fatal for the vanquished. But it is still necessary to control aggression as wounds sustained in a fight can go septic, and are a major cause of death. Rank is established by fighting but, thereafter, the hierarchy minimises conflict. Each monkey can recognise the others and predict the most likely outcome of an interaction with any individual; the strong gain access to priority items and the weak avoid being further weakened. There is little time-wasting for all members of the troop who can get on with other tasks.

In general, adult males hold the highest rank, above juvenile males and all females, though there have been exceptions from time to time. Each matriline has a rank too, established by fighting among the females at the time the group was formed, and all members of one matriline outrank all members of a subordinate one. Within a matriline things are a little more complicated. As juvenile females reach puberty they rise in rank over their older sisters, but not their mothers, so that the most dominant female is the mother, followed by her youngest daughter and any of her offspring, then the next daughter and her offspring, and so on. Young males benefit from their mother's rank until they are adolescent, but when they disperse to a new group they have to establish

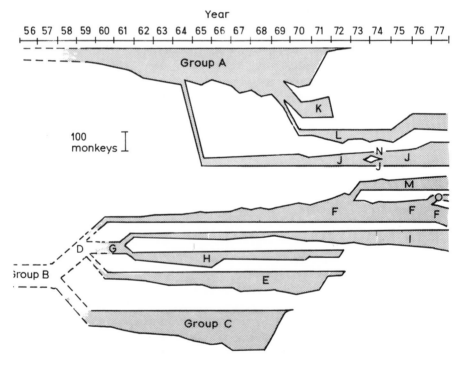

Figure 1 *A summary history of the social groups on the island. The depth of the shaded areas represents the number of animals in the groups, which split when they become too large*

themselves slowly and carefully, which often means a few years at the bottom of the troop hierarchy.

Many years of careful and systematic observation have given us a clearer idea of what the monkeys do and how they organise their social life. But still we do not know what causes them to behave the way they do. General social structure is consistent across the six groups on the island but, despite this and the ecological uniformity of the island, we still find significant differences in the amounts of friendship and fighting from troop to troop. There is much to learn and we are advancing slowly and steadily. The evolutionary approach to biology has provoked many people into quite wild flights of theoretical fancy. The monkeys of Cayo provide an ideal set-up to check out some of these ideas, and results are coming in. Group splitting, shown in Figure 1, allowed Diane Chepko-Sade to ask whether the monkeys are aware of their relations; she found that they were (*New Scientist*, vol. 82).

We are looking at microevolution by checking on blood groups and enzymes on a regular basis. We are relating male rank to reproductive success using the same blood groups; we can definitely exclude certain animals from certain matings and can often be reasonably sure who was the father of which infant. Female reproductive strategies ought to change with age, as the loss of future offspring becomes less important and we are looking for evidence of these changes.

As we pass our 40th anniversary the need for primate colonies has become, if anything, more pressing. The conditions that led Carpenter and the others to set up the colony on Cayo have not improved. Habitats are still being destroyed and species are still under threat. Biological stations such as Cayo Santiago can relieve the pressure imposed by medical research on wild populations. They can also enhance our understanding of the nature and development of primate behaviour and provide a valuable yardstick against which to measure studies in the lab and observations on wild animals.

C

7

Rutting on Rhum

ROGER LEWIN
16 November 1978

In order to test theories about the functional value of different behaviour patterns, it is necessary to observe a population long enough to see which individuals are most successful in terms of the number of offspring they leave. That is the object of a Cambridge University study of the red deer on a remote Scottish island.

Fiona Guinness lives on the Isle of Rhum for 10 months of the year, a pretty isolated existence by any standards. Her home is a small stone cottage – a bothy – that rests on a gentle rise close to where Kilmory Glen opens into the narrow stretch of sea that separates Rhum from the Isle of Skye, its famous Coolins snatching jaggedly at the skyline. Every five or six days Fiona sets out from the bothy, notebook, tape recorder and binoculars in hand, to complete a census of her nearest neighbours – nearly 350 red deer (*Cervus elaphus*) which roam over an area of about 10 square miles. She does not simply count the animals. She knows every individual by sight (they each have a name or a code number) and she notes down what they are eating, which other individuals they are with, what the weather is like at the time she sees them, and many other detailed observations. Fiona is not some fanatical eccentric who has dedicated her life to communing with the red deer of Rhum. She is the crucial member of a team of researchers who is exploiting the Isle of Rhum's natural laboratory in order to watch the cutting edge of natural selection in action.*

---

* The entire Rhum red deer team consists of Steve Albon, Michael Appleby, Tim Clutton-Brock, Rosemary Cockerill, Robert Gibson, Fiona Guinness, Marion Hall and Michael Rice. The project is being successful only because it is very much a team effort, different people's studies intermeshing closely together to create something approaching a total picture of the animals' behaviour patterns.

For most of the year, Fiona is accompanied by a couple of colleagues from the research team, each of them checking particularly on some aspects of the animals' behaviour – mother–infant interactions, for instance, or male dominance hierarchies during the island's bitter winter months. But, come October, the small bothy fills to bursting point as the Cambridge University biologists who are manning the project descend on the island. It is the rutting season when, in evolutionary terms, prizes are lost and won. Females seek out areas where the grass is lushest and the stags compete with each other for "ownership" of harems. Within the space of a couple of frantically active weeks next season's offspring will have been sired and conceived: some stags will have fathered several offspring, some none; some hinds will be pregnant, some will have been in too poor a condition to conceive.

Roary, who lost half an antler after a vigorous encounter with Boss

Broadly speaking, those animals which at the end of their lives have left behind many children can be counted more successful in the context of natural selection than those whose children are few. If the factors that determine an individual's reproductive success during its lifetime are inherited by its offspring, then they in their turn will be the winners of the annual October rut. As generation passes to generation, the descendants of the successful individuals will become more and more numerous in the population. It does not take any complicated mathematics to demonstrate that if one animal has even a small competitive edge over another, it will, given enough time, enough passage of generations, enjoy a considerably greater genetic representation in some distant future population. This is the stuff of natural selection.

When they were planning the project, Tim Clutton-Brock, Steve Albon and their colleagues said to themselves, "Let's assume that the red deer behave in the way they do now as a result of millenia of evolutionary selection". This being so, if Clutton-Brock and his team can discern what it is that determines the so-called life-time reproductive success of stags and hinds (the size of stags' antlers, for instance, and the nutrient quality of hinds' home range), they can truly claim to be watching the mechanics of natural selection in action. This is the objective of the Rhum red deer project.

Rhum is the stub end of a once massive volcano, or perhaps there were volcanic twins – opinions differ. It is now one of a scattering of stunningly beautiful mountainous islands in the Inner Hebrides, off Scotland's west coast. Through human intervention in crofting agriculture during the past couple of centuries, Rhum is now virtually naked of trees. The Nature Conservancy took over the island in 1958 and turned it into an open-air laboratory, the red deer experiment being just one of many fascinating projects there. (Another adventurous project, incidentally, is replanting many areas with forests such as covered the island long ago.)

Fiona Guinness first went to the island nine years ago, before the Cambridge team had established itself. She started observing the animals in a fairly casual way while she was engaged on another project. It is only in the past five years that she has been applying the detailed recording system that is the true backbone of the new project. The researchers of Rhum are not the only people watching large mammals with an academic eye: lions, hunting dogs, baboons, chimpanzees – all have their devotees taking down in painful detail every item of their daily activity. The truly exciting point about the deer project, though, is that very soon it will have collected important behavioural activity on a large group of animals from conception to postmortem decay. This sort of information is crucial if one is to answer evolutionary questions in behavioural terms.

For 11 months of the year there is a rough segregation of sexes in the red deer, segregation, that is, of mature animals. Hinds, perhaps with their new calves and yearlings, move around together in search of the best food they can find. The groups, the members of which tend to be related to each other more or less distantly, usually occupy particular areas rather than there being a total free-for-all over the whole range. Inevitably, some parts of the island are better nutritionally than others and this means that calves born to mothers who live in one of the richer regions have a better start in life. One of the things that the study has shown is that lighter-than-average calves have a higher-than-average chance of dying in early infancy.

Towards the end of the summer, when the hinds have been building up reserves that will serve them – and possibly a growing embryo – through the lean winter, they move to traditional rutting grounds. Once again, hind families appear to have a regular claim on certain areas, and these too vary in the quality of food they offer. The hinds that command the best quality winter and summer quarters are usually those most likely to produce good calves.

Meanwhile the stags, which for most of the year wander around together in more or less peaceful coexistence, making a living on rather poorer quality vegetation than that upon which the hinds have been feasting, suddenly become a good deal less tolerant of each other. They then make for the greens where the hinds are congregating. The competition commences gently at first, but pretty soon it is in deadly earnest, with the ultimate booby prize being death. Each stag's aim – if one can express in such terms a pattern of behaviour which is to a large degree innate – is to secure for himself as large a harem as possible in which he will be able to inseminate as many hinds as are in oestrous.

Stags usually live to around 11 or 12 years of age, during only half of which time at the very most will they have been rutting success-fully. They are practically mature at five years but only rarely hold any sort of harem before they are six. The peak years are roughly between seven and 10, with a very short geriatric period before they die. Hinds, by contrast, have a longer reproductive life, beginning at three and going on until 14 and longer.

The key element that determines whether or not a stag has a successful rutting season is his fighting ability. In order to take charge of a group of hinds in the first place a stag may have to fight a rival for the right to them if he cannot find any stray ones. Once he owns a harem he will certainly be challenged over them repeatedly. If he beats all-comers, he will, with luck, finish up with a harem of perhaps 20 or more animals and will be able to mate with those hinds that come into oestrous.

But rutting is a very exhausting business, involving assiduous attention to the hinds and flinging oneself full-bloodedly into fights with would-be usurpers. More characteristically, it includes continual sentry duty and regular roaring. Part of the physiological preparation for the rut in stags is the development of enormous neck muscles and changes in the voice box. The stags depend on their massive necks in the strenuous antler to antler fights, and with their enlarged vocal apparatus they produce deep and massive roars that echo around the hills day and night, especially during the second two weeks in October, the peak of the rut. The stags also spend a lot of time marking their territory, by urinating and by thrashing vegetation with the head, leaving behind a scent from glands near their eyes.

Because of the preoccupations of the rut, stags with harems usually spend less than half an hour a day feeding, as compared with the normal six or more hours. Stags, therefore, have to rely principally on their fat reserves for energy during the rut and so they have to be careful not to exhaust themselves assembling a big harem before the hinds are ready to come into oestrous. The crucial time to hold a harem is when they are ready to be fertilised.

Fighting is a serious business for the stags: not only does it consume an inordinate amount of energy but it is also extremely dangerous. Every year a quarter of the stags suffer some sort of minor injury and more than one in 20 incurs permanent damage. The two contestants lock antlers and push against each other until one is forced backwards, a result that may be achieved either by advantage of weight or, more cunningly, by fancy footwork and exploitation of sloping terrain – brains as well as brute force are important. One stag near to Kilmory, an eight-year-old called Roary, employs even more cunning. His antlers are unusually lacking in branches, a structure which is to some extent disadvantageous in that he is less able to get a firm grip on his opponent's headgear, but it is a disadvantage which Roary turns around because he quite clearly tries to use his long unbranched points as lethal rapiers.

In a recent fight with Boss, a highly successful stag, Roary repeatedly attempted to slip his rapier-like antlers through his adversary's more conventional multi-branched equipment and at one point he even attempted to spear Boss in the flank. Boss was too quick for him though and blocked the thrust. In the end Roary lost more than the fight with Boss, he also retreated with a broken antler.

Roary's misfortune in that encounter emphasises the fact that fights should not be entered into lightly. Moreover, in addition to the threat of physical damage, there is the even more likely danger

(Above) A stag 'marks' a post by rubbing scent-releasing glands against it

(Left) A young male (a staggie) grooms himself

that during the bout young stags, geriatric stags, and any other stags around will take advantage of the owner's preoccupation and will steal the temporarily abandoned hinds. So-called sneaky rutters (young stags up to the age of five) are a perpetual irritation to harem holders as they hang around the group in the hope of picking up a stray female. The mature stags expend a great deal of energy chasing these youngsters away.

How to minimise the costly exercise of strenuous battle? The answer is that one advertises one's fighting prowess in a roaring contest, the idea being for, say, a harem holder to discourage a challenger from entering into a fight. When a stag sees a group of females that he would like to include in his own harem, he begins

(Left) A hind with a young calf

(Below) Stags with harems occasionally take a rest from their duties; they do not often roar while seated

roaring at the incumbent, perhaps at a rate of four a minute. Now, if the harem holder replies with seven roars to the minute, the challenger will more than likely take his bravado elsewhere. If, however, the roaring contest is pretty even, the challenger will approach closer until eventually he is perhaps 10 yards from his potential opponent. At this point the two animals go into what is called a parallel walk: they walk tensely shoulder to shoulder, separated by between three and 15 yards, still roaring, but by now away from each other. If the challenger wishes to push things further he lowers his head and a fight begins. Challengers frequently choose a strategic withdrawal during the parallel walk; harem holders rarely do – they have too much to lose.

The research project has demonstrated clearly that an animal's roaring characteristics (principally the rate) correlate very closely indeed with its fighting ability. One must, therefore, conclude that evolution has produced a genuine system for advertising the likely outcome of a potentially dangerous encounter. Cheating – through bluff – apparently does not work; at least, evolutionary forces do not seem to have favoured it here and there are theoretical reasons why this should be so generally.

Stags are about half as big again as hinds – just as you would expect in a species in which males hold harems (gorillas and hamadryas baboons are other examples of polygynous animals). But what is somewhat surprising is the relatively small difference in life-time reproductive success between stags who are good fighters and those that are not. Suppose a stag holds a big harem, say 20 animals, through the oestrous period. Of these, only half will be mature hinds, 20 per cent of these females will not breed that season, 20 per cent of the offspring will die in early infancy and a further 10 per cent will not survive the winter. The stag will, therefore, sire five surviving offspring that year, and given that peak rutting lasts typically for just four years, the probable maximum is around 20.† This is a surprisingly low jackpot for individuals who are apparently willing to stake so much in the first place. The gamble, however, must be worth it, otherwise natural selection would have pushed the stags along another path of siring.

During the rut, the activity of the whole research team increases. Fiona Guinness, for instance, steps up her census to once a day. She notes down which hinds are in whose harem throughout the period and so when the calves are born 235 days later she knows pretty well who fathered them. This information is crucial to determining the life-time reproductive success of the stags, to see how many come close to that theoretical maximum of 20 surviving children.

The theoretical limit of fecundity for females is one calf a year from the age of three until she dies at around 14 or so. There is an enormous physiological strain in giving birth and rearing infants year by year and only hinds in continual peak condition can manage this. Usually they have a season off every third year or so. The

† It is now thought that the maximum number of offspring for a stag is 25 and for a hind, 12.

difference in descendants between champion hinds, who merely have to have access to good feeding grounds, and super successful stags, who have to fight every inch of the way is, therefore, surprisingly small. But, for one stag competing against another, any advantage is in the end significant when measured in the balance of natural selection.

Once the frenetic period of the rut is over, the deer are rapidly plunged into the exigencies of winter. Stags that have rutted hard – perhaps successfully, perhaps not – will be seriously depleted of body stores and this can be the death knell for the older ones. In April, the stags shed their antlers, soon to be crunched between the deer's jaws for the precious calcium they contain. May and June bring abundant food supplies and newborn calves. The stags begin to grow a new intact set of fighting gear, at first covered with a sensitive velvet. By late July the antler growth is complete and the now lifeless velvet hangs in festoons. August passes, and soon September is at an end – the rutting season returns and the knife edge of natural selection once again hangs over the stags. Some will succeed. Some will be less lucky.

8

Red deer raise more sons than daughters

'MONITOR'

19 February 1981

As expected, red deer put more effort into raising male calves than female. But surprisingly, this does not mean they produce fewer of them.

Red deer and other mammals invest more heavily in male than female offspring. But, contrary to prediction, they do not raise fewer males, as would be expected from the observation that males can produce more offspring in a lifetime than females. Scientists in Tim Clutton-Brock's Large Animal Research Group at Cambridge University have reached this conclusion after collecting copious evidence, especially from the famous red deer on the Isle of Rhum in Scotland.

A red deer (*Cervus elaphas*) hind can expect to give birth to a maximum of 12 calves during her lifetime. A stag, by contrast, could do more than twice as well and father 25 or more calves, because although the reproductive life span of the stag is shorter, he lives in a polygynous society and, if successful, will mate with a number of hinds each season. Parents should, if they obey evolutionary principles, divide their total investment in the next generation equally between the sexes, which would mean producing fewer of the sex that is more costly. The Rhum Red Deer Project has been collecting data on identified animals for many years now and Clutton-Brock, with Steve Albon and Fiona Guinness, has analysed the data to see if there is any evidence that hinds invest more heavily in male calves (*Nature*, vol. 289, p. 487). They find a great deal of support for this idea.

Male calves spend longer in the womb, are heavier at birth and probably gain weight faster. They suck twice as often as females and each suckling bout lasts longer. That all this is indeed a greater drain on the hind's resources is shown by the fact that hinds which successfully raised a male calf through the winter were far less likely to conceive again the next summer than those who had reared a female. Those that did conceive after a male calf, gave birth later than after a female. Hinds also calved later in the season after rearing a male than after a female; late calving reflects the mother's poor condition and is linked to higher mortality in the calf.

All this is a good thing for the male. His success in the breeding game is crucially dependent on his size and physical condition and his full-grown size depends more on growth during the first year than at any other time, so the male that does well in his first year will probably do well when mature.

The hinds invest more heavily in each male offspring, so they ought to produce fewer of them. But this does not seem to be the case. In fact, they seem to do just the opposite and there are 57 males to 43 females at birth. By one year, six months after weaning, the ratio has dropped to 53 males to 47 females, but still the imbalance exists. This is a challenge to theory, which says that there should, if anything, be fewer males than females.

Clutton-Brock, Albon and Guinness suggest two ways in which their results can be reconciled with theory. It may be that mothers who allow daughters, but not sons, to stay on the home range with them have to put up with competition for food and hence are effectively investing further in the daughters. But unrelated red deer hinds share overlapping home ranges, so that a mother who did exclude her daughters would gain little in the way of resources. Alternatively, the ratio of males to females *at conception* may be fixed at unity. If this is true, and if individual sons require greater parental investment from their mothers, natural selection could favour parents that give a greater proportion of total investment to sons. Against this, there is some evidence that sex ratios at conception are not fixed at unity.

So the case for greater investment in the sex with the higher potential breeding success – males – is well substantiated. But the enigma of why *total* investment in each sex is not equal remains.

9

Tracking birds through space and time

PAUL HARVEY and PAUL GREENWOOD
10 April 1980

After 30 years, Oxford University's study of great tits in Wytham Wood began to reveal some answers to questions about long-term population structure and dispersal.

We often hear anecdotes about birds that nest in the same suburban garden year after year. But, surprisingly, we know very little about where the young fledglings go to breed or where the parents were born. If we knew how individual birds move around from place to place we might gain an insight into other questions, such as whether birds commit incest and, if so, with whom? Present-day evolutionary biologists are particularly concerned with such problems because they deal with population structure. Since the 1930s, population genetics has developed very much as a theoretical science, partly because we cannot yet answer questions like those posed above. Only when we get realistic estimates of the various components of population structure can we expect the elusive synthesis of ethology, ecology and population genetics to become a reality. But that synthesis can be achieved only through long-term studies of known individuals carried out by teams of dedicated observers. One such project is beginning to bear fruit 30 years from its conception.

The great tit (*Parus major*) is a monogamous, woodland bird which, in England, produces one clutch a year. The adult male defends a feeding territory and keeps other great tits away. Offspring from one breeding season normally produce their own young the following summer. One advantage of studying this species is that the birds will use artificial nest boxes in preference to natural nesting sites. In 1912, the Dutch ecologist G. Wolda began a study in Holland which H. N. Kluijver continued. After visiting their site,

David Lack initiated a parallel project in 1947 at Oxford University's Wytham Estate. More recently, Christopher Perrins, director of the Edward Grey Institute at Oxford, managed the project and extended it to include over 320 hectares of woodland saturated with about 950 nest boxes. Between 130 and 270 pairs breed in the study area each year.

Great tits like to make their homes in nest boxes, which makes them easy to study

In Wytham, the birds start breeding in late spring and the young leave the nest in early summer. The female can be ringed and identified while she incubates the eggs and the male is best caught as he feeds the young. Nestlings are ringed prior to fledging. With the ringing data one can construct detailed genealogies and learn how birds move about the woods. With the enthusiastic cooperation of Chris Perrins, we used the Oxford records to answer some of the questions about population structure outlined above.

The adults are remarkably faithful to one site. If a pair produces a successful brood one year and both adult birds survive the following winter, they can usually be found in the same territory the next year. If, however, their young are taken by predators, they are likely to breed elsewhere the following year. Even so, the birds are very conservative and move to nearby territories rather than to a completely different area of the wood. Occasionally, predators kill the brood early in the season. When this happens, the pair immediately attempts to breed again elsewhere within the same territory. (After all, the other territories are occupied, so where else could they go?) But even if the second attempt to raise a brood is successful, the birds seem to remember their earlier failure; they are likely to use a different nesting territory the following year. Divorces do happen, in which instance it is almost invariably the female who moves away to a different territory where she breeds with a new mate the next season. Males seem to be particularly faithful to their old territories, staying on them whenever possible.

Because it is the males who establish territories, we might ask whether a son takes over his father's territory. The answer is yes, but only if the father has died; otherwise, the sons seem to nest as near as possible to the territory they were born in. Females, however, move farther and tend to end up about seven territories from where they were born. Using a computer simulation produced by Mike Webber, a postdoctoral worker at Oxford, we were able to show that the females do not move randomly around the wood; if they did they would nest about nine or 10 territories from their birthplace.

The sex difference in natal dispersal, where daughters move farther than sons, might occur because it prevents inbreeding. We surveyed the literature and found that rates of inbreeding and levels of inbreeding depression (the drop in breeding success that occurs when close relatives mate) had never been measured in a natural population of animals. Even so, it is frequently assumed that inbreeding is harmful, but that offspring disperse too far to make incestuous matings likely. Given the low degree of dispersal in the great tit and many other vertebrate species, we would question both these assumptions.

We did a complete search of a sample of 1000 matings from parents of known genealogy, and uncovered 10 brother–sister and seven mother–son incestuous matings. In view of the site tenacity of adults and the sex differences in natal dispersal, we were not surprised that there were no father–daughter matings in our sample. Indeed, using a model developed by Michael Bulmer, the Oxford biomathematician, we were able to show that the frequency of inbreeding between the various types of relative was consistent with the overall pattern of dispersal in the two sexes. In addition, when we looked at how far the inbreeding birds had moved, we found that among the brother–sister pairings, the males had moved further than normal and females less, while among the mother–son matings, the sons had moved shorter distances than normal outbreeding male birds.

Even more exciting results came when we looked at the breeding data from those 17 incestuous matings. Nestling mortality was nearly double that of outbreeding pairs. If the tendency to disperse can be inherited, the fact that it also prevents inbreeding could well be a potent factor that influences the way that fledgling great tits move away from the territory they were born in. Although no other study of birds has been able to provide enough data to investigate inbreeding depression, we can find many reported cases of sex differences in dispersal, generally with the female moving farther than the male.

So far we have concentrated on those factors which favour dispersal. However, many field studies of animal and plant populations have revealed remarkably low levels of dispersal. Explanations for such philopatry (meaning love of the fatherland) are normally couched in general ecological terms. For example, some people argue that an animal is particularly well adapted to the place where it is born, either genetically or – among vertebrates – because it has learned where food supplies are and how the predators behave. Another possibility is that an animal may find it easier to establish a territory when a relative is setting one up next door. A series of models (based on the concepts of "inclusive fitness" and the "evolutionary stable strategy" – see Chapter 22, p. 61 and Chapter 24, p. 171) has been developed which predicts that this might occur because aggressive interactions, such as those involved in setting-up and defending a territory, should be less severe between close relatives.

*A great tit (*Parus major*) takes off*

Where males are the territory defenders, are there reproductive advantages to male kin nesting in adjacent territories? An early laying date has, in the past, been correlated with reproductive success in the great tit. We have discovered that when related males nest in contiguous territories, their mates begin to lay earlier than expected. However, even with the data already available, we have been unable to demonstrate greater success, in terms of number of offspring fledged, in neighbours who are related. Perhaps another 30 years' work will provide the answers.

Nevertheless, findings such as those we have outlined here highlight the importance of long-term genealogical data for testing current theories. They should also lead to the development of new-finely tuned models which represent the real world more accurately than those presently used by population geneticists.

10

Bird spotting by radar

ERIC EASTWOOD

12 April 1962

New and better aircraft in the postwar era meant new and better radar to track them. To the radar engineer's annoyance, birds showed up well on the screen and ornithologists were able to use the technique to make new discoveries about bird behaviour.

By the end of the Second World War the coasts of the British Isles were defended by a formidable array of radar stations which had gradually been built and commissioned by the three services. These various radar chains covered a very wide frequency spectrum, from the 20 megahertz (MHz) band of the original CH (Chain, Home) system through 200 MHz of the coastal CHLs (Chain, Home, Low) and inland GCIs (Ground-control, Intercept) to the S and X band microwave sets of the "low cover" and fire-control systems. The total of some 200 stations provided a wealth of uncorrelated radar data which were vital to the defence forces during the war but which possesed little relevance to civil air traffic control in peace time. The inevitable result was that only a few of these stations continued into the postwar era, and these were soon to be replaced by new military radars that were better able to cope with the changing air threat arising from the introduction of high-speed jet aircraft.

The new aircraft of the 1950 era were able to fly both higher and faster than those of wartime; moreover, the target cross-sections which they presented to the ground radars were greatly reduced, while their ability to screen themselves by the use of electronic jammers was much increased. It was the sum of these factors that compelled the substantial improvements in ground radar equipment and this has resulted in the development of transmitters whose powers have been increased by a factor of 10, so bringing them into the megawatt range. Receiver sensitivity was similarly improved by at least 12 decibels, while the need for high resolving power and narrow beamwidths compelled the adoption of frequencies in the microwave band in order that ½° beams could be achieved by the use of rotatable aerial arrays of reasonable apertures. The performance of the high-powered radar of the 1950s was such that a single seagull could be detected at a range of 70 miles.

It was the introduction of radars of this type into operational use that called attention to the presence of echoes upon the plan-position radar displays other than of the aircraft it was desired to observe. The occasional occurrence of such so-called spurious echoes, or "angels", had been observed upon the wartime radars and identification of some of these echoes with birds had been made by Dr David Lack during his work for the Army Operational Research Group. The RAF station at Happisburgh had similarly identified flights of geese upon its 1½-metre radar as early as 1942. It did not at first seem possible, however, that the masses of angels often seen upon the PPIs (plan-position indicators) of the new high-powered radars could be explained in terms of birds. Nevertheless, detailed observation by Dr Sutter in Switzerland and by Lack and other workers in Britain, including the Marconi Research Laboratory, has proved that a substantial proportion of angel echoes are indeed attributable to birds. The immediate result of this work has been to establish radar as a powerful tool in ornithological research; it has also revealed the remarkable amount of bird activity which takes place by night.

The obscuration of the PPI by angel echoes could become an operational hazard when the radar is used for military or civil air traffic control, unless special techniques are employed to reduce their density. In order that suitable methods could be devised for eliminating the echoes, special investigations were undertaken at the Marconi Research Station on Bushy Hill in Essex which had as their objective the better understanding of the angel phenomenon.

In this work, the diurnal and seasonal variation in intensity of angel echoes has been studied and the flow lines that the angels followed have been observed by a recording method similar to the use of time-lapse photography for the display of slowly changing phenomena, such as the growth of flowers. The PPI display is fitted with a special cine camera whose film-transport mechanism is linked to the aerial of the radar. The normal revolution speed of the aerial is 4 rpm, so that a complete 360° scan of the aerial occupies 15 seconds and is recorded upon one frame of the cine film. When such a film is projected at the normal speed of 16 or 25 frames per second, the movements of the birds are made readily apparent and speeds of flight may be measured.

These radar records, which have now been maintained since 1958, have confirmed and extended the work of Lack and Tedd performed in 1957 at an RAF station in Norfolk, which established the general correspondence between angel occurrences off the East Coast and the migratory movements of birds across the North Sea. The Bushy Hill records are now permitting a detailed analysis to be made of all bird movements around the southeast corner of England. This is a region in which bird movements are probably more complex than for any other position in the world, by reason of the situation of the British Isles in relation to the European land mass and the proximity of this location to the paths of the major movements of birds that occur over the whole of this vast territory in the spring and autumn. The bird movements in this region are greatly affected by meteorological factors such as wind and temperature; radar is helping to unravel these effects and is also contributing to a better understanding of the problem of how birds navigate between their breeding grounds and winter feeding regions.

The speeded film method of analysis of bird flight has made a particularly interesting contribution to the study of the flocking and roosting habits of starlings. The radar films show that the usual method of dispersal of starlings from their roost at sunrise takes place in the form of a succession of waves of birds, separated in time by an interval of approximately 3 minutes, and which move outwards from the roost to the feeding grounds in the form of a circular wave. This type of movement is illustrated in Figure 1, where a number of ring dispersals of birds may be seen. The fact that the starling exhibits this characteristic roosting behaviour has allowed it to be easily identified upon the radar PPI and so has permitted an intensive study of a bird species by means of radar. A radar record of starling activity within an area of 14 000 square kilometres centred on the Bushy Hill station has now been maintained since 1958; this wealth of observational detail has led to many new conclusions upon the various factors which influence the roosting and flight habits of the starling.

The identification of other species of birds seen upon the radar is not so easy, although factors such as type of echo and speed of movement can provide useful clues. In particular, altitude of flight might prove to be a useful identifying feature of a species. The Bushy Hill radar system is served by a high-power height-finding radar which by "nodding" – scanning in a vertical direction – allows the

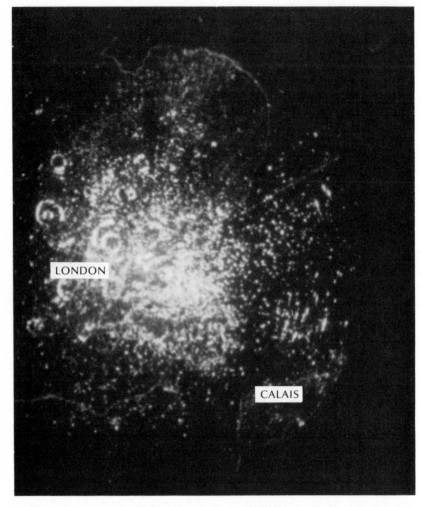

Figure 1 *Radar ring angels observed on 25 July 1959 as starlings took off from their roosts at sunrise in London and the southeast*

altitude of the birds to be measured. Studies are now in progress to determine whether certain altitude bands tend to be characteristic of different types of bird movements. Figure 2 shows a composite picture of a PPI display and that of the associated nodding height-finder. On the PPI display a selected angel echo is surrounded by a small circular strobe marker. The corresponding echo is identified upon the height display by the small arrowhead and, in this case, is found to correspond to an altitude of 3000 m. From measurements

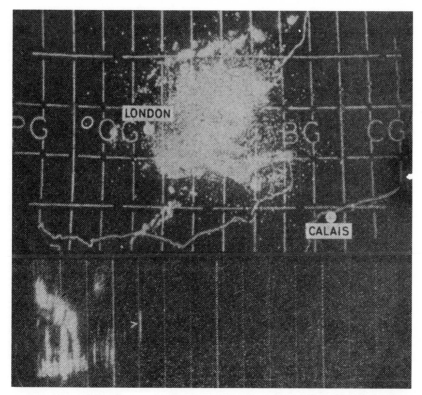

Figure 2 *Clouds and angels on a radar plan-position indicator (PPI) and its associated height display. A flock of birds is marked with a ring in the PPI; the same flock is indicated with an arrow-head in the height display, which shows it to be at 3000 metres*

of a large number of angels, it appears that the most common altitude for migratory flight is about 1000 m. Bird echoes at 4000 m are by no means uncommon, and occasional records have been obtained up to 6000 m. It will be clear that radar is the ideal bird-watching tool for providing information upon birds in flight, and that certain parameters of bird flight which have hitherto been inaccessible to the bird watcher can readily be measured by the radar technique.

Two particularly impressive examples of the power of the radar tool in ornithological research are provided by Figures 3 and 4, which are concerned with the movement of birds at a sea-breeze front and the vesper flight of swifts respectively. On 20 June, 1960, there was a cloudless sky over the southeast of England; it was a

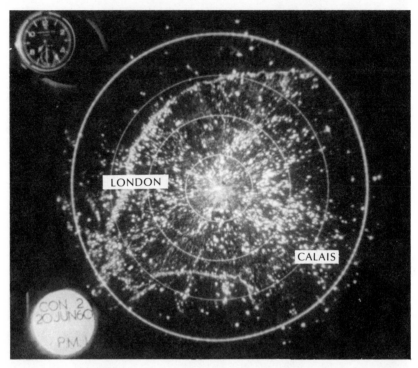

Figure 3 *The PPI shows birds soaring at a sea-breeze front*

beautiful summer day. About midday a line front was observed to develop on the radar along the north coast of the Thames Estuary and along the Kent-Sussex coastal strip. These frontal movements developed during the afternoon and moved inland at a speed of 6 knots (kn). The situation at 1830 hours in the evening is illustrated in Figure 3. The total linear extent of the front is in the order of 160 km; its leading edge is very sharp indeed. These radar lines show the surface of demarcation between the moist air penetrating from the sea and the warmer, drier air over the land. Calculations show that such a front would produce a refractive index change that might cause the front to be just visible to the radar. Closer inspection of the photograph will show the mass of small echoes on the trailing edge of the front which, by use of the speeded film method,

Figure 4 *(a) and (b) Radar record of the vesper flight of swifts. (a) PPI at sunset, 2100 h. (b) PPI at 2220 h. The swifts are airborne and have drifted out to sea on a light breeze*

a)

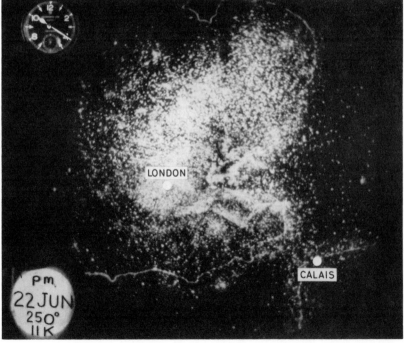

b)

can be shown to possess speeds of movement away from the front in the order of 27 kn. This speed corresponds to that of a bird in flight and so there can be little doubt the the main reason for the visibility of the front to the radar is due to the concentration of soaring birds at the front; probably both seagulls and swifts are involved. It should be emphasized that this phenomenon took place in a perfectly clear, cloudless sky.

A similar radar observation of a sea-breeze front was made from Bushy Hill on 22 June, 1959, under identical weather conditions. The front dispersed completely at sunset, and Figure 4a shows the PPI display at 2100 hours: the radial white line to the northwest is in fact produced by radio noise from the setting sun. With the onset of night the bird activity remained comparatively slight until 2220 hours when the density of angels increased rapidly and produced the PPI picture shown at Figure 4b. The nature of this PPI picture shows that a mass of small birds had taken to the air over the whole of the area covered by the PPI. The concentrations in the neighbourhood of towns are to be noted, especially those on the south coast and on the coast of France in the region of Boulogne and Calais. When observed by the speeded film method this transition from a quiet PPI to the widespread activity shown at Figure 4b is quite breathtaking.

Detailed examination of these records by Lack and ourselves has convinced us that this phenomenon is to be explained in terms of the vesper flight of the swift. It has long been known that it is a habit of the male swift to take to the air in the late evenings of high summer but it has not been known whether these birds spend the night on the wing. This radar evidence provides striking confirmatory proof of the theory that these vesper flights of the swift, as they are termed, do indeed take place over widespread regions of the country and that a substantial proportion of the birds spend the night on the wing. Since there is a tendency for a greater concentration of nesting swifts in the region of towns, the effect of these vesper flights is to produce a strengthening of the radar bird echoes at points corresponding to these locations. The presence of a slight breeze off the land drifts the birds gradually out to sea and so produces the apparently displaced coast lines and town "blobs" lying in the sea which are to be seen at Figure 4b.

Radar bird watching is, of course, no replacement for the normal methods of visual observing. It is a powerful new tool which must be used in association with visual watchers but these few illustrations will serve to show that a great wealth of new information on bird movements is likely to result from its application.

11

Biotelemetry: listening in to wildlife

DAVID MACDONALD and
CHARLES AMLANER
19 February 1981

Radio tracking and biotelemetry enable the modern biologist not only to observe animals but to monitor their physiology. Creatures that would be frightened by a human observer go about their daily business unaffected by the presence of a radio transmitter.

For millenia men have struggled to interpret the tracks and signs left by other animals. Under the tutelage of men such as Niko Tinbergen, fieldcraft has grown from a woodman's art to a biologist's science and yet, still, a gust of wind can erase a line of tracks in the sand and carry the unwelcome odour of ethology to the quarry. Before the days of radio tracking the very simplest question – where does the animal go? – has been largely unanswerable, especially for the secretive mammals. But since the early 1960s, when William Cochran and R. Lord (at the Cedar Creek field station in Minnesota) published the circuitry for the first radio transmitter appropriate for use on wild animals, it has become increasingly easy to discover not only where animals spend their time but also, at least to some extent, what they do there and with whom they do it. Indeed, radio tracking helps even in finding one's subject. Brian Bertram (now curator of mammals at London Zoo) scoured the Serengeti for leopards, but thwarted by their camouflage he spotted only four in 30 months. Yet once he equipped a leopard with a radio collar he could find the marked animal at will. What is more, once the leopard wore a radio Bertram could observe it for longer periods because he did not have to stick so close to the animal and hence reduced the risk of "spooking" it.

Radio tracking has now transformed field studies and promises to provide answers to a host of outstanding biological questions. What is more, the next two decades should see a similar revolution

wrought by the more sophisticated cousin of radio tracking – biotelemetry. While tracking simply allows a biologist to locate an animal, biotelemetry implies transmission of information on any measured (metred) biological variable. Already this has included information on an animal's behaviour and physiology and on the nature of its surroundings.

In practical terms, radio tracking and biotelemetry involve fitting a miniature transmitter to the subject using a collar, or a harness, or perhaps by implantation inside the animal's body. The miniature transmitter emits a signal – commonly VHF (>100 MHz) and often in pulses – via an antenna which may be a loop (around the neck, and serving also as a collar), a whip (a straight wire), or a coil of copper wire around a ferrite rod. A receiving antenna with reciprocal properties to the transmitting antenna picks up the signals at a range of several kilometres or more. Then a beat frequency oscillator (BFO) translates the received signal into sound. The signal may be converted by a demodulator (a "black box" which changes the output of the receiver) into a form appropriate for direct computer processing, storage on tape, or visual display. Some receiving antennas are directional; the strength of their signals (the volume of the perceived bleep) is at a maximum or minimum when the antenna points at, or is perpendicular to, the incoming beam of radio waves. Thus the biologist can tell the direction of the signal by its volume. In principle, using triangulation, he should be able to pinpoint his animal's position, or, in the jargon, to take a "fix" on it.

The biologist is, however, cursed by topographic error. The direction from which he receives the signal is not necessarily the same as that from which it left the animal because radio signals bounce along the sides of valleys, ricochet off rock faces, filter between detached houses and become polarised by stands of conifer trees, all of which would lead the textbook tracker in pursuit of spurious fixes. People who fail to check their triangulations while still in the field sometimes find, when the time for analysis comes, that the

---

Figure 1 *Ian Lindsay's method of observing badgers. A recorder close to the sett monitors their immediate comings and goings: 1 is a car battery; 2 an automatic timing device; 3 a tape recorder; and 4 the receiver (with its aerial attached which picks up the signal from the antenna on the badger). Badgers that wandered away from the sett could be followed by car, or on foot*

home range of their strictly terrestrial animal encompasses sub-stantial areas of the Atlantic Ocean. The tracker must remain a "fieldman", knowing every detail of the landscape and its radio idiosyncrasies. To some extent the tracker can reduce errors by adjusting the frequency: radio signals of lower frequency (30–50 MHz) are less easily deflected than those of high frequency (200–500 MHz). However, it is the licensing authorities rather than the biologist's preferences that normally determine which frequencies are available.

Transmitters on foxes around Oxford enabled Macdonald and colleagues not only to observe where the animals went, but also what they did

Since 1972 we have equipped red foxes (*Vulpes vulpes*) in and around Oxford with various designs of radio collars, of which the most recent is the two-stage transmitter. When used to the full this equipment can do more than gather simple information on each fox's location: it enables the tracker to know where his animal is, to predict where it may go next and, hence, how best to stalk or ambush it for observations. Radio tracking alone tells an impover-ished story but it has spawned a whole subsidiary technology to facilitate night-time observations. We encapsulated bulbs of phosphorescent tritium (H$_3$), called betalights, in clear plastic and mounted them on the fox's collar so that it could be spotted at night from more than 250 m with 10 × 50 binoculars. Collars were fitted with individually patterned reflective markers and, perhaps most importantly, a motion-sensitive (mercury) switch incorporated into the transmitter circuitry enabled the tracker to tell whether a fox was stationary or active.

By radio tracking, we found one vixen in the same place within an orchard on three consecutive nights; through careful stalking and observations with infrared binoculars she was watched on each occasion. One night she was eating windfall apples, on another she was foraging for earthworms, on the third she was sleeping. Radio tracking alone would have told us that she was in the same place each night. The only drawback to such observation is that the modern naturalist must prowl the darkness entwined like a human Christmas tree in a tangle of leads and bedecked with the trinkets of science.

That radio tracking has come of age was amply demonstrated at a symposium held jointly by the Zoological Society of London and the Mammal Society, organised by Chris Cheeseman and Ron Mitson from the Ministry of Agriculture, Fisheries and Food (MAFF).

Among the most ingenious of all devices is that described by Graham Hirons of the Game Conservancy in Hampshire and Ray Owen of the University of Maine who work on the mysterious "roding" behaviour of the male woodcock (*Scolopax rusticola*), a bird whose camouflage renders it almost invisible to the naked eye. Roding is a crepuscular display during which the male flies above the tree tops making a strange croaking noise. After sunset, however, the roding woodcocks are no more than black dots against the sky, so even with the birds equipped with radio transmitters, Hirons and Owen could find out little about their displays except that they took place over an area larger than 100 hectares. But then they fitted a bird with a transmitter (on a backpack) in which a resistor had been replaced by a thermistor, which was so arranged that it lay snugly under the bird's wing. When the woodcock began its nocturnal flight the thermistor chilled rapidly and altered the resistance in the circuit, which in turn reduced the bleep rate of the received signal (a form of information coding known as pulse interval modulation, PIM).

Hirons and Owen found that birds which secured more matings also roded more. Although the details of why some male woodcock

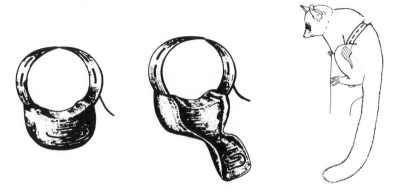

Figure 2 *In West Africa, Charles-Dominique fitted bushbabies with waist-bands with a circuit that shorted when the animals urinated in order to mark their territory. Hence he studied their territorial behaviour*

rode more than others still remain undiscovered, it is clearly an expensive feat; a few days of roding could cost the performer 25 per cent of his bodyweight.

In 1977, Pierre Charles-Dominique from Brunoy, near Paris, described a comparably ingenious modification of an externally mounted transmitter. He was studying a small nocturnal animal, Allen's bushbaby (*Galago senegalensis*), in the tropical forests of West Africa. Each bushbaby was equipped with a waistband from which hung a transmitter. To the rear of the transmitter protruded a lip in which lay the exposed ends of two wires connected to the transmitter circuitry. When the bushbaby urinated, drops of fluid shorted the wires, allowing current to shunt through a lower resistance loop so that the transmitter pulsed at a different rate until the urine evaporated. The transmitters were delicately fashioned to accommodate the anatomical differences between male and female bushbabies and so Charles-Dominique was able to study urine marking in both sexes. In this way, at the same time as plotting the animals' movements, he discovered that urine marking was most frequent at the perimeter of bushbabies' territories.

Signal changes have shown their potential in other, equally diverse contexts. T. R. Loughlin of the National Marine Fisheries Service, Washington, DC, has fitted 164-MHz transmitters to sea otters (*Enhydra lutris*). Initially, Loughlin was puzzled that on one hand sea otters were reputed to feed only in the late afternoon, while on the other they have a high metabolic rate, eat up to 25 per cent of their bodyweight each day and have little subcutaneous fat. How could they tolerate fasting by night?

Sea otters spend much of their time floating in beds of kelp close to the shore. They gather shellfish from among the kelp or from the rocks down to 20 metres. By watching radio-collared animals by day, Loughlin realised that because VHF signals were greatly attenuated in sea water, he could identify, from the signal alone, three types of behaviour that were distinguished by signal strength and constancy. The categories of behaviour were resting, active but not feeding, and foraging. By recording these same categories at night, Loughlin found that this was when sea otters did about 45 per cent of their feeding. Loughlin buckled the transmitter harness around his otters with metal staples which corroded and fell off after the study ended.

Biotelemetering features of the environment can complement information transmitted simultaneously from the animal, as Robert Kenwood of the Institute of Terrestrial Ecology has shown by his work on grey quirrels (*Sciurus carolinensis*). The collars he used on these squirrels were descended from the original design of Cochran and Lord and modified to include a bead thermistor.

Kenwood fitted one such temperature-sensitive device to the squirrel and put another in its drey. When the squirrel went into its drey the temperature around the thermistor rose, so Kenwood was able to tell when the animal was in its nest. Similarly, Ian Lindsay of the Animal Behaviour Research Group at Oxford has been studying the times at which badgers (*Meles meles*) emerge from their setts. He fitted badgers with radio collars sensitive to motion. There were too few people available to monitor the badgers' journeys to and from their setts, while at the same time following individuals who strayed. With an automatic timing device which we describe in the *Handbook of Biotelemetry and Radio Tracking*, this logistical impasse was overcome. The timing box activates a radio receiver and tape recorder for selected periods and prescribed intervals. Subsequently the recorded signal could be interpreted in terms of the badger's activity and presence or absence from the sett. In the meantime, Ian Lindsay could follow other radio-tagged badgers elsewhere, or occasionally even go to bed!

Radio transmitters have also yielded some insight into predator–prey relations. The first "mortality transmitter" was reported by Dave Mech from the US Forest Wildlife Service in Minnesota as early as 1967 – in that case the pulse rate changed when a snowshoe hare (*Lepus americanus*) was killed by a fox and its temperature dropped. Mech also discovered that wolves and their prey, white-tailed deer (*Odocosilus virginianus*), were partitioning the woodlands of Minnesota. During the winter months the prey gathered in aggregations at so-called deer yards. Radio tracking of deer and wolf revealed that these deer yards were slotted into the boundary zones between the adjacent territories of wolf packs.

In another predation study Robert Kenwood, together with Vidar Marestrom and Mats Karlbom from Uppsala, Sweden, assessed the ecological and economic importance of attacks by goshawks (*Accipiter gentilis*) on game species on a sporting estate in Sweden. Transmitters were mounted on the hawks' tail feathers and thus were shed at the next moult. These researchers neatly worked out how to estimate the numbers of goshawks by employing the Lincoln Index method. They knew how many goshawks they had tagged, and could estimate the total number by seeing what *proportion* of all the goshawks that they subsequently counted, were fitted with transmitters.

However, the researchers wasted a lot of time seeking out radio-tagged hawks which they thought had killed pheasants, but had not. Robert Kenwood solved the problem by judicious placement of a mercury signal which gives a fast pulse when the tail is horizontal (in flight), slow pulse with the tail vertical (perched), and an alternating pulse as the bird arches it back to pull at food.

To monitor a heart rate of a vertebrate by biotelemetry may seem achievement enough, but comparable studies of invertebrates need acutely sensitive transmitters. A crab's heart is little more than a thin-walled bag of muscle, producing an electrical signal that is smaller than those from a mouse's heart by orders of magnitude. Nevertheless, T. G. Wolcott of North Carolina State University has produced a transmitter whose signal pulse rate varies with the electrical activity of the myocardium (heart muscle) or cardiac ganglia (nerve control) of tiny ghost crabs. Wolcott knew what a crab (*Cancer magister*) was doing from its heart rate; it rose from 30 to 150 beats/minute with 5 seconds of beginning activity and fell again to resting level within 2 minutes of the crab's standing still. In another invention, Wolcott incorporated a light emitting diode (LED) into a transmitter's circuitry to aid noctural observations. A colony of ghost crabs, each shouldering a turret from which various hues of flashing light beamed across the sands, evoke an image in keeping with its name!

In general, radio tracking should be viewed as a way of minimising a biologist's intrusion on his subject's life. A radio transmitter

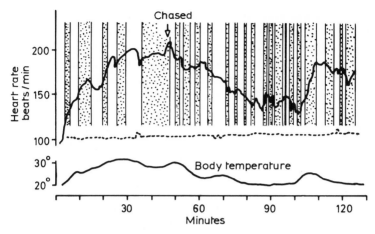

Figure 3 *T. G. Wolcott fitted ghost crabs with transmitters to measure variations in heart rate and body temperature as the animal runs and rests*

that is so cumbersome and awkward that it interferes with its bearer's behaviour is not only to be deplored as inhumane, but dismissed as useless for sensible scientific purposes. In the early days of biotelemetry, research workers tried to minimise the influence of the transmitters by reducing their size and weight – conventionally to less than 5 per cent of the animal's body weight. Nowadays, transmitters are so small that they can even be implanted inside the animal. It is at least arguable that this is more humane: where once the animal might have been confined to a small cage and adorned with trailing wires, it may now be subjected to only a minor operation and then released. Objectively, too, such operations seem to do little harm. In one study, Harvey Smith of the US Forest Service in Connecticut implanted transmitters into the peritoneal cavities (abdomens) of mice of the genus *Peromyseus* and found there were no long-term deleterious effects: indeed 37 per cent of females with implanted transmitters produced young, compared with 33 per cent of females without transmitters. In addition, implanted transmitters are often necessary to study the physiology of animals in the field; indeed, such studies are now breaking down the artificial barriers between physiology and ecology.

At the meeting at London Zoo the staggering implications of this point were emphasised by the work of Pat Butler and Tony Woakes from Birmingham University. Their study of diving birds is notable not only for the new insight which it brings but also because it turns conventional wisdom unceremoniously upon its head. A generation of students learned that when a duck dived its heart rate slowed (bradycardia) and that some of its blood vessels constricted so as to reduce the flow to the animal's surface, both reflexes being designed to conserve oxygen. This notion stemmed from experiments wherein a duck strapped to a seesaw had its head forcibly dunked in a bucket of water. The fact that several of the ducks used were dabbling species – not divers – must have made them less willing collaborators during these rather unpleasant experiments. During these enforced immersions, the concentration of oxygen in the blood went down and this reduction, together with a build-up of lactic acid (which is produced when sugars are broken down in the muscles without oxygen), led to the belief that diving ducks submerged with inadequate oxygen supplies to sustain their aquatic travels and supplemented their energy supplies by anaerobic respiration. The heart rate takes, however, about 60 seconds to

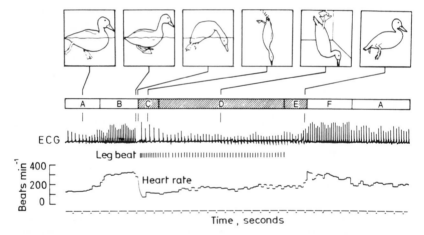

Figure 4 *Butler and Woakes's studies of tufted ducks challenge accepted ideas on diving reflexes. Biotelemetry shows that as they prepare to dive and to resurface their heart races (shown in ECG and heart rate traces); but when they are submerged and feeding, the heart rate slows. In the bar, A is the time spent on the surface; B, of cardiac acceleration before submersion; C, descent; D, feeding on the bottom; E, surfacing; and F, the period of cardiac acceleration before surfacing. The drawings are traced from a ciné-film of a duck in a glass-sided tank*

reach its lowest level in response to falling oxygen tension in the arteries (a reflex mediated by the carotid body). So, as a way of conserving oxygen, this mechanism has an element of closing the stable door after the horse has bolted. Indeed, many natural dives did not last as long as 60 seconds. When Butler and Woakes implanted a telemeter which transmitted an electrocardiograph (ECG) signal from unrestrained pochard and tufted duck (both diving species), they found that when the birds dived voluntarily there was no bradycardia and that the supply of oxygen in the blood was generally adequate without resorting to anaerobic respiration.

The same authors now seem poised similarly to demolish the dogma of flight physiology. With the aid of a tame barnacle goose, implanted with heart and respiratory rate telemeters, Butler and Woakes are recording from the bird as it flaps along amiably beside their moving vehicle. Already it seems that the electrical impulses from the muscles of a bird flying in a wind tunnel differ from one flying naturally.

Norbet Smith, from Northeastern State University Oklahoma and his collaborators Diana Worth and Lisa Causby, have bio-telemetered the heart rates of free-ranging rabbits (*Sylvilagus floridanus*) and woodchucks (*Marmota monax*). When disturbed both exhibit a so-called fear bradycardia. Following the Butler and Woakes principle, when woodchucks are exposed to the same disturbance in captivity, their heart rate increases (tachycardia) – exactly the opposite response!

Radio biotelemetry has been largely transformed from a cause of concern for animal welfare to a powerful tool for welfare and conservation. For example, in an investigation of the welfare of domestic hens, Ian Duncan and John Filshie of the Agricultural Research Council's Poultry Research Centre in Edinburgh noted that some "flighty" strains of hen exhibit more avoidance and panic to visual stimuli than "placid" strains. However, the heart rate of ostensibly placid birds rose almost as much, and took longer to recover than that of the apparently more "flighty" birds. Similarly, Nigel Ball and Charles Amlaner of Oxford Animal Behaviour Research Group emphasised how appearances can be deceptive as they monitored the heart rate of a gull at the approach of a walking human "predator". Without any discernible change in the gull's appearance its heart rate shot up from 240 to nearly 600 beats per minute when the predator came within about 5 metres. Indeed, the shape of the heart-rate response curve seemed to vary with the risk to which the gull was exposed.

At the London meeting, Bob Stebbins of the Institute of Terrestrial Ecology, Monkswood, provided a compelling case for the value of radio-tracking studies to the conservation of the rare greater horseshoe bat (*Rhinolophus ferrum-equinum*), whose numbers today are only one per cent of those a century ago. In the past two years, Stebbins has glued transmitters to the fur on the backs of 27 of these bats. Almost instantly, his ideas of their behaviour and, hence, of the requirements for their conservation, were revolutionised. Apart from discovering the bats' predilection for sleeping in stately homes, he located vital roosts which must be protected and feeding areas over old pastures. Stebbins emphasised how the accidental loss of even one bat would be intolerable with such a rare species. So care had to be taken in using radio transmitters. For example, they could not be used on heavily pregnant female bats for fear of critically altering their wing loading.

Many field studies have an applied side and distinctions between pure and applied work seem largely unhelpful; studies of the nightly rambles of badgers and foxes contribute to the control of bovine tuberculosis and rabies, knowledge of the movements of leopards and elephants aid park authorities on deciding the limits of game reserves. Fishery scentists too are using comparable techniques, based on sonar, to monitor shoals.

All in all, there is an appealing irony in radio tracking and biotelemetry; a high technology that might at first have seemed antipathetic to traditional fieldcraft is helping today's biologists to realise the dreams of generations of naturalists and to study wild animals more closely than ever before. With their advantages of keen senses and swiftness of limb, our subjects still have the edge on us, but radio tracking and biotelemetry are levelling the odds.

PART THREE

A Sense of Purpose: Behaviour and Evolution

For Western biology, the 20th century has been the age of neo-Darwinism: the concept of evolution by natural selection, as proposed by Charles Darwin, augmented by the genetic theories that began with Gregor Mendel and were finally capped, in the 1950s, by the elucidation of the structure of the gene itself. According to this view of evolution, new characteristics may arise through chance mutations of the genes and be preserved or thrown out according to whether or not they increase an individual's fitness – loosely speaking, its chances of surviving to breed. Mutations that did have been termed "adaptive". This generalisation applied to behaviour as much as to structure. Ethologists reasoned that where most members of a population were regularly engaged in a particular form of behaviour, there must be sound evolutionary reasons for it, however pointless it might seem to the casual observer. Ever since, they have applied themselves to the task of accounting for the vast diversity of animal behaviour in terms of its contribution to fitness.

Being ourselves social animals, we take a close interest in the social lives of others. What are the costs and benefits of living alone, in a small family group or in a large flock or troop? Early attempts to answer this question, like the article by M. R. A. Chance (p. 89), were based on rather perfunctory observations and tended to come up with simple conclusions – large groups offer protection against predators, for example. More recent studies like those of Tim Clutton-Brock (p. 98) show that the simple answer is often not the only one, even if it is correct as far as it goes. Extremely subtle differences in feeding habits may account for the surprising variation that exists between closely related species.

The functional approach, with its emphasis on the transmission of genes to future generations, inevitably leads to a preoccupation with sex. Yet again, animals display numerous different strategies for ensuring that their genes have a safe passage to the future. Some associate briefly for copulation, produce large numbers of potential offspring and invest no effort whatever in their upbringing; others entice a member of the opposite sex into lifelong partnership, both parties devoting themselves to their few and precious young ones. Between these extremes, animals exploit numerous other possibilities. It is the ethologist's difficult task to work out why what is right for, say, a gorilla does not suit a chimpanzee (p. 109).

Despite their apparently submissive role, females often turn out to wield a surprising amount of power in their relationships with males. They can accept one and spurn another, and the spurned will have failed in the game of life if he can find no one to welcome his advances (p. 118). It is unusual for females to dominate males in a more direct sense but not unknown: the female jacana, a bird of Central America, keeps not one but several males in subjection (p. 124). It seems to be worth her while to do so, so why have other species not evolved the same system?

Perhaps some of the most difficult problems for ethologists concern nature's most aesthetically successful accomplishments. The peacock's tail is a classic example. What could a peacock want with such a gaudy and cumbersome appendage? The answer seems to be that he needs it to dazzle his wives but need he really go to such lengths? The intricate and musical songs of many bird species pose similar problems. But like other forms of display animals practise, song can be a threat as much as a means of attracting a mate. Only detailed study of the species in question can reveal whether the bird is more interested in defending a territory (like the great tit, p. 132) or winning a bride (like the sedge warbler, p. 139).

12

What makes monkeys sociable?

M. R. A. CHANCE
5 March 1959

In the 1950s knowledge of the behaviour of non-human primates was slight but even then scientists were looking for functional explanations of the variety in social organisation that was apparent among ,monkeys studied up to that date. They were perhaps a little too eager to make comparisons between the behaviour of monkeys and that of humans.

Man has been defined as a speaking and tool-making primate. Judging by the amount of our lives that is usually devoted to these activities in modern societies, were are certainly that type of creature but these activities are also social. We are, however, less aware of this side of our behaviour and have shown little desire to study normal individual and social behaviour – except as a means to an end, such as warding off the difficulties arising in different social groups, or in order to cure mental illness. This is because we have not found out how to see ourselves objectively without the disturbing influence of social tradition and so do not know that behaviour could be other than what we think it to be now.

Tradition created us and is one of the means by which we have evolved from our animal past, but much of this past is denied recognition by tradition. We are thus prevented from seeing that our animal heritage is with us, cloying our creativeness as well as furnishing us with the wherewithal to live in the way we have evolved.

We are, therefore, in need of information about the properties of our behaviour in order to know what our real potentialities are. But existing knowledge of our behaviour is couched in traditional terms. How do we overcome this dilemma?

One way of edging round this difficulty is to study the behaviour of our near zoological relatives. At present, however, behaviour studies of the lower primates are not far advanced. Nevertheless, sufficient has emerged from the few existing pieces of research work for us to be able to distinguish certain basic elements at least in outline and to see how the social bond, which is so strongly developed in man, might have originated.

Unfortunately, the primates anatomically closest to us, the apes, which are classified as the Hominoidea, live in tropical and sub-tropical wooded country, so that it is difficult to observe their behaviour. Few detailed observations have been made on apes, with the exception of the gibbon, but it seems clear that they do not form large groups. We must, therefore, be content at this stage with observations on the Anthropoidea, or monkeys. Some types of monkeys spend much time in groups and the average size of the groups can be used as a rough measure of their sociability.

As it happens, sociability measured in this way seems to be closely related to the animals' habits; those monkeys which spend a lot of time in open country seem to form large social groups, whereas those which live in forests tend to have small and loosely co-ordinated groups. This can be seen from Table 1.

The theory of evolution has provided biology with the most powerful heuristic method of thought. It helps biologists to fix their attention to the central issues of the problem by demanding that all biological phenomena should be related eventually to the selective advantages of adaptation and competitive existence. So here we simply start by asking what advantage do monkeys foraging in open territory derive from living in large groups? What is the origin of this "social bond"?

The simplest answer is that when feline predators are about, monkeys are safer in numbers. Observations on baboons in Africa, which form closely marked social groups, point to this conclusion. It is confirmed by the observations of C. R. Carpenter who saw a young howler, a South American monkey, which was isolated from its group, attacked and bitten by an ocelot. The young monkey cried out and received protection from three males of the group.

The fact that gibbons do not appear to form social groups, but live in families, can be explained by the same argument. Gibbons live in tree tops, where they are relatively free from predators, although they can be attacked by leopards, which climb trees easily. But gibbons have a unique method of escape by swinging – brachiat-ing – from the slender branch of one tree to that of another. Leopards and other tree-climbing cats would have difficulty in following them over such arboreal pathways.

Table 1 Habitat and Group Structure in Different Species of Primate

| Species | Habitat | | | Group structure | | | | | |
	Habit	Country	No. in group	Males	Females	Heterosexual groups socionomic index*	Lone males	Bachelor bands
Gibbon (Hylobates agalis)	Brachiating; structurally adapted for life in trees	Thailand	Less than 6	1	1	100	(+)	–
Spider monkey (Ateles geoffroyi)	Walks along branches; lives mostly in the trees	Central & South America	5–30	8	15	160	(+)	–
Howler monkey (Alouatta palliata)				3	8	230	+	–
Hamadryas baboon (Papio hamadryas)	Lives almost exclusively in open country	Central Africa	25–100	1	12	?		
Rhesus (Macaca mulatta)	Forages in open country and in the forests	India	Up to 150	6	22	500	+	+
Japanese monkey (Macaca fuscata)	Inhabits partly wooded and open mountainous country	Japan	25–440	2	4	Small group 2 males 4 females	+	+ (Separate from but associated with the heterosexual group)

Notes: * Number of females for every 100 males.
+ = present, – = absent, (+) = inferred.

The spider monkey, a long-limbed animal native in tropical America, seems to occupy an intermediate position between the howler and the gibbon. These are relatively large animals and usually move on all fours. But they are able, like the gibbon, to escape from danger by brachiating from tree to tree. They have a looser form of social group than the howler, which cannot brachiate.

Although the adaptive features of living in groups are not exhausted by the protection afforded against predators, particularly in open country, it seems probable that this was the adaptive feature of their behaviour which led many species to develop the faculty to form large groups.

These groups show a definite structure, with a predominance of females and young. As an adolescent male matures, he either takes his place in the main group – from which, at the same time, an older male is often expelled in order to maintain the ratio of males to females – or he is himself expelled.

These excluded males represent an interesting feature of the social organisation of monkeys. It is difficult to follow their fate while they remain solitary, and some presumably fall victim to predators. But observations of excluded howlers show that some of them quite soon become associated with another group. They gradually become acquainted with one member of it and are ultimately incorporated into the new group. Again, this may mean that some other male is thrown out of it so that the ratio of the sexes is maintained.

The rhesus macaque shows a more exaggerated form of this behaviour. Its heterosexual groups are large and contain a high proportion of females so that a large number of males are expelled as the adolescents mature. Some of these excluded males join together to form bachelor bands which exist separately from heterosexual groups. They contain old males which have retired from dominant positions in the breeding hierarchy and young, excluded males.

The same social structure exists in the Japanese macaque, which has been closely observed on a peninsula set aside as a reserve for the monkeys. The stability of the bachelor bands is remarkable: they continue to exist even though bachelor and heterosexual groups have to come close together in order to visit the feeding site. The most likely explanation is that the males of the bachelor band have an attraction for one another.

When most kinds of animal are brought into close proximity in this way by having to visit feeding or breeding grounds, the resulting encounters stimulate both aggressive impulses and the tendency to flee. Aggressive encounters are especially intense between mammalian males during the breeding season, when the conflict is frequently resolved by fights – as, for example, in deer. These fights merely serve to establish a rank order between the males, which then take it in turn – the dominant stag starting – to run with the herd of hinds.

Rhesus monkey mothers, daughters and young. The young adult males are banished from the main social group and live in bachelor bands

A rank order between the males of a social group has been found to exist for all species of social monkey that have been studied: for example, the Japanese and rhesus macaques, the hamadryas baboon, South American howler and spider monkeys. Their order determines the priority of access to food and mates, to sites for sleeping etc, and imparts freedom of movement within the society as a whole. More significant, however, is the fact that there are never fights between male members of the monkey groups, except on the rare occasions when the dominant male or overlord is displaced by another monkey. This sometimes occurs by the simple retirement of the overlord, or after a period of assertive behaviour on the part of the monkey which will ultimately displace him. Only during such episodes is the stability of the rank order disturbed. Monkeys of lower order seem to be well aware of their place in the community.

Around this constant core of males, the breeding relations are organized in different ways according to the species of monkey. For example, male baboons possess harems, the size of which is proportional to the rank of the male. A group of Japanese monkeys was studied over a period of several years, during which time the group varied from 160 to 440. During the whole of this period, six adult males formed the core of the society. These males followed strict rules which led them to mate with females of their own harem, or any other female, provided she was not attached to the harem of other males of the core. A female belonging to a harem does not permit herself to mate with a male other than her overlord for, if another male is permitted to transgress, the retaliation of the overlord is directed at her and not at the transgressing male. This indicates the extent to which the male rank order has suppressed hostility between its members. Its effectiveness is all the more remarkable, since the male monkeys are faced persistently with intense sexual provocation.

Female macaques and baboons are sexually receptive for approximately nine out of the 28 days of their reproductive cycle, so that in a group of monkeys where two or more adult females are present, the males will be in the presence of a sexually active female for more than half the time. And in larger groups there will be continual sexual provocation – a situation found nowhere else in the animal kingdom, except for a two-month interval during the mating season of the Pribilof seal. The extent to which this preoccupies the male seal can be judged from the fact that he does not feed at all during these two months but lives on his fat, spending his time defending his territory and his harem.

The only explanation of the monkeys' ability to stick together in spite of this provocation seems to be the specific social attraction between members of the male sex which also accounts for the formation of bachelor bands. Sociability is, therefore, more than an aggregation of a number of individuals in the face of external threats.

During the observation of a colony of rhesus macaque monkeys at the London Zoo in 1953, I was startled to see a female being aggressive towards her prospective consort. As I watched her, I saw him counter-threaten her several times until she was attacked and bitten, after which she paid him no further attention.

Subsequently, I found that other females courted in the same way, but with important differences. The threat exerted by a courting female towards her prospective male consort *was usually combined with submissive gestures*, and in this case it did not lead to counter-attack but served to arouse his interest in her.

Here was a clue to the social bond uniting the males. It has always been clear that the subordinate males are aware of the *potential* threat from those higher in rank. This threat has presumably acquired an ambivalent quality for monkeys – it attracts as well as repels them just as the female's display of modified aggressiveness can attract a male. Any evaluation of their social behaviour must take into account this strange ambivalence.

The next step in my observations of the rhesus macaques was to study the evidence of attraction between males under varying conditions – when they were grouped without consort females, when only the overlord had a consort and when the two leading males had a consort each. One way of measuring the degree of attraction is to study the spacing of the males: the closer the spacing, the greater the attraction. From these studies it appeared that *potential* threats – as conveyed by the awareness of the higher status of another adult male – act primarily as an attraction to subordinate males. The presence of consort females, however, brings about mild *overt* threats down the rank order. These engender a repellent influence between the males in the breeding hierarchy, which is shown by an increase in the distance between them.

Very recently, Hans Kummer of the Basle Zoo has completed a detailed analysis of the behaviour of a family of hamadryas baboons living in an open enclosure. His observations confirm the conclusions just outlined concerning the ambivalent response to threat. Concerning young monkeys which have left the mother but are not yet mature, Kummer wrote:

During expressions of fear, the frightened individual does not remain where it was threatened. Either it flees from the cause of its fear or it seeks out an animal of the highest possible rank. When fear is intense the latter invariably happens. But in most cases the highest ranking animal within range of the dispute is itself the cause of the fear. This does not significantly change the behaviour of the frightened animal. It seeks out the highest ranking of the animals present though this individual has himself been the cause of its fear.

After a subordinate monkey has run to the high ranking monkey, his behaviour shows unmistakable signs of conflict. Ambivalent postures, screeches etc, show his tendency to run away, yet he continues to sit nearby.

The young monkey, from a very early age, habitually seeks out its mother, and for much of the time she clutches it to her and suckles it. Contrary to what one might expect, however, Harry Harlow has shown by the aid of model mothers that the mother is regarded primarily as a refuge rather than as a source of food.

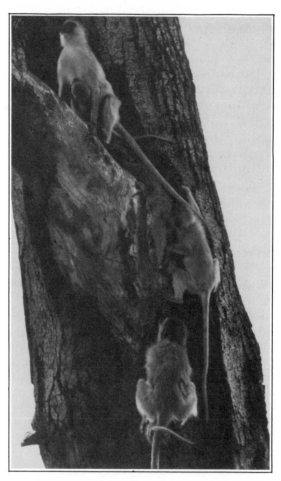

Monkey mothers carry their infants almost all the time. These are vervet monkeys in Tanzania

Hans Kummer gives us a glimpse of the final link in the succession of events which forge the social bond between the growing monkey and the other members of the colony when he writes:

> It seems that with advancing age the elements of behaviour of the young in the relationship with their mother do not vanish *but project themselves upon ever higher ranking individuals.*

Much study will be needed before we completely understand the dominance of one monkey over another, or the behaviour of the subordinate monkey towards the dominant one. But it seems clear enough that a successful social relationship between them depends on the achievement, by the subordinate, of a balance in his attitude to the superior. He is subjected to conflicting influences, the most important of which is the ambivalent quality of threat already mentioned. The monkey's awareness of potential threat from high ranking animals is further balanced by the tendency for subordinate monkeys to seek out the high-ranking animals during social conflict.

The tendency to run to a leader is usually explained by the protection that he affords but that is not necessarily the true cause of this type of behaviour which, in monkeys at least, is a compulsive switching of attention towards animals occupying a particular social position.

History is replete with examples of man's tendency to line up behind an intimidating leader – the emergence of such a tendency in Adolf Hitler's Germany being only one dramatic and ultimately disastrous demonstration of this behaviour, which we can now see may have a strong instinctive component. The tendency must be clearly distinguished from democratic methods of delegating authority to a leader for a social task.

Can it not be that this instinctive tendency often takes control of our behaviour without our being aware of what has happened? If so, to become aware of our instinctive social motivation will help us to lead more congenial lives with greater personal freedom. The study of the social monkeys will help us to distinguish the elements requiring our attention.

13

Why do animals live in groups?

TIM CLUTTON-BROCK

11 July 1974

In the 15 years after article 12 was written, a wealth of data accumulated on the social lives of many species. It was (and is still) not possible to formulate a comprehensive theory to explain differences in group structure among similar species, but at least in some cases the advantages of living in groups seemed clear.

Outside my window, four blackbirds and 15 starlings feed busily in the spring grass. The starlings are tightly packed and evenly spaced. The flock moves forward steadily, occasionally lifting off and moving to a new feeding site. If disturbed, the birds will bunch and escape together. In contrast, the four blackbirds are widely distributed and forage independently. If one strays too close to another, it is likely to be attacked, and if danger threatens they will fly off singly, taking different directions.

Most families of higher animals contain species which show similar contrasts in grouping patterns. Some species of deer and antelope are solitary or live in pairs, often defending small territories, while others live in vast, loosely-knit herds. Among both the carnivores and the primates, a number of species live in large groups of stable or semistable membership while others are solitary or live in permanently mated pairs. In many species, group size and composition vary between populations and, in some, between times of year, though a model pattern of grouping can usually be identified.

At the moment, the adaptive significance of these differences in grouping is poorly understood. Biologists have proposed a wide variety of suggestions about the advantages of living in groups. These suggestions fall into three obvious categories: advantages related to defence or avoidance of predators, those related to finding or handling food, and those concerned with reproduction. I have been studying the behaviour of red deer in an attempt to test some of these ideas. Before describing my own findings, I should like first to discuss more fully other observations of group behaviour.

Grouping may increase the chances that a predator will be noticed before it can attack and may allow the animals to defend themselves effectively. For instance, many bird species bunch together when a hawk is seen, a reaction which may deter the predator from swooping into the group. Similarly, a variety of ungulate species bunch at the approach of carnivores and, if attacked, some show cooperative defence. Grouping may also help to reduce the risk of predation in other ways. Making the assumption that a large group of animals is no more attractive to a predator than a small group or a solitary animal. William Hamilton (at the Imperial College Field Station) argues that an individual will decrease its chance of being attacked by putting itself close to another animal and that this could cause natural selection to favour individuals that aggregate.

Group living may also provide a reservoir of knowledge about the places where predators are to be expected and about the correct tactics to be taken in the event of an attack. Experiments with chimpanzees show how effectively an animal that has seen a predator in a particular spot can later communicate its presence to other group members when reintroduced to the same place (see p. 244).

The second advantage of grouping relates to obtaining food. Arthur Ghent in Toronto describes how sawfly larvae collect in clusters to feed on pine needles. The larvae attracted to sites where one of their number has already cut through the cuticle of the needle and are able to feed there without having to open the cuticle for themselves. The large African carnivores that hunt cooperatively provide another example and in several species the size of hunting groups varies according to the size of the prey. In these animals, group living may also act as a form of evolutionary insurance policy. Working with lions in East Africa, George Schaller argues that membership of the pride helps to tide individuals over times when they are sick, wounded or pregnant and cannot hunt efficiently for themselves.

Feeding in groups may also help animals to find food. Individuals can learn where to search by watching other group members – a type of social facilitation known by ethologists as "local enhancement". This idea has been experimentally tested by John Krebs. He introduced great tits singly and in small groups into an aviary where

Zebra and other grazing animals may group together to protect themselves from large carnivores

food (mealworms) was hidden. The experiments showed that birds feeding in a flock of four were more likely to find the food during a 15-minute interval than when feeding on their own or with one other individual. Krebs and his colleagues also discovered that after one bird had found food in a particular hiding place, the other birds in the flock concentrated their attention in the same general area and on the same kind of hiding place.

Local enhancement may operate at a higher level, too. Birds that nest in colonies or roost in flocks may learn where to search for food from the behaviour of incoming or outgoing groups and the colony or roost may act as a clearing house for information about the whereabouts of good feeding sites. Grouping may also help animals to utilise their feeding areas efficiently by reducing the chance that an individual will search for food in an area that has been depleted.

Finally, grouping may enhance reproductive efficiency. In species where individuals are widely dispersed or where the sexes are normally segregated, grouping can be particularly important in helping animals in breeding condition to locate suitable partners. Through social facilitation, it may also promote the development of breeding synchrony. This may prevent predator populations from accumulating or, in species where breeding is cued to food availa-

Wading birds feed in large flocks only if mutual interference does not affect their ability to find prey

bility, ensure that the decision to start breeding is based on information drawn from a wide area.

Perhaps the only safe generalisations that can be made about the functions of grouping is that, in any one species, they are likely to multiply, and that they differ between species. The important problem is to identify the key factors responsible for differences between particular species. Three kinds of evidence have been commonly used: experimental evidence; correlations between environmental factors and inter-specific differences in grouping patterns; and similar correlations at an intra-specific level.

Experimental evidence is scarce. Moreover, in higher animals, it is virtually impossible to design an experiment that will test the effect of differences in grouping on fertility and mortality. Experiments have concentrated so far on investigating the effects of grouping on other aspects of behaviour. This has the disadvantage that the effects that are demonstrated may not be those through which selection is operating to maintain group living. For example, Kreb's experiments with great tits showed that individuals feeding in groups were more likely to find food than those feeding alone. In a wild population, this effect might be counterbalanced by increased competition resulting from flocking, and the selective advantage of grouping could really lie, say, in increased efficiency of predator detection.

Several studies have related inter-specific differences in grouping patterns to gross ecological differences. One problem with this approach is that correlations at this level are likely to be weak because different species may adapt to similar environmental situations in different ways. For example, black-headed gulls protect their nests and chicks from predators during the breeding season by an elaborate system of colonial defence. Kittiwakes, nesting on cliff ledges that are inaccessible to predators, lack many of these defensive tactics. In addition, different species may evolve similar traits in different ecological situations. A final problem with this approach is that, where correlations exist, they often provide little information about the underlying mechanisms.

After reviewing the available information on a wide array of bird species, the late David Lack was able to show that few insectivorous species feed in flocks, while many seed-eating species do so. A variety of different reasons could be responsible for this association and the survey offered little indication as to which were important. These problems can be partly resolved by investigating the ecological correlates of differences in grouping between pairs or small groups of phylogenetically and ecologically similar species.

Recent studies have adopted this approach with wading birds, zebras, hyenas and colobine monkeys. In most cases, differences in groupings have been found to be associated with differences in food dispersion or in feeding method. Often, these are of an extremely subtle nature, as John Goss-Custard's studies of wading birds show. Knot, dunlin and black-tailed godwit usually feed in closely packed flocks which, among dunlin and knot, may include several hundred birds. In contrast, oystercatcher, curlew and redshank feed singly or in small flocks and individuals are widely spaced. Both groups of species feed on the open mudflats, taking inter-tidal worms and small molluscs. However, they forage in different ways. The first group feed by rapid pecking or continual sifting and probably locate their prey by touch. The second make distinct pecks at the mud's surface and may use visual clues to detect food. Since only a small proportion of the prey population is visible from the surface, the number of prey available to these species is probably lower than for the first group.

Mutual interference may discourage the formation of large, tightly packed flocks in these animals, either because the birds' weight on the surface of the mud causes the prey to withdraw from sight or because the leading birds remove a large proportion of the available prey. By making a captive redshank walk along a Perspex tunnel placed on the mud's surface, Goss-Custard showed that a bird's weight does cause a substantial proportion of the available prey to disappear from the surface. The effect would presumably be unimportant among species in the first group, which do not use visual cues to detect their prey.

Further support for this suggestion comes from observations of variation in flocking and feeding behaviour within species. When feeding at night, redshank (normally a single feeder) collect in large compact flocks, feeding principally on *Hydrobia*, a small gastropod mollusc that does not react to disturbance on the mud's surface. While they are in their nocturnal groups, the birds change their feeding methods and they collect their food by sweeping their bills from side to side, probably detecting the prey by touch. Though this argument explained why oystercatcher, curlew and redshank normally feed in small, dispersed groups, it did not show why the other species aggregate in large, compact flocks. Goss-Custard argues that this helps them to minimise predation.

Like redshank, many animal species show marked temporal or spatial variation in grouping patterns. They offer a particularly good opportunity to study the effects of different ecological factors on grouping, since these effects can be more readily identified than in inter-specific comparisons where a large number of differences are inevitably involved. My current studies with Fiona Guinness of red deer on the Isle of Rhum (see p. 50) show how wide a range of factors can influence grouping patterns. Red deer hinds on the island aggregate in matriarchal groups, typically consisting of a matriarch, her mature daughters and their dependent offspring of both sexes. Groups are not stable and individuals associate temporarily with members of other groups or move away to feed alone. However, except in cases where a daughter has left her natal group and founded a matriarchy of her own, hinds associate much more frequently with members of their own matriarchy than with other animals.

Each group has a well-defined range which is totally overlapped by those of surrounding groups. Within these ranges, groups associate in temporary aggregations which may contain 50 or more animals. Weather has an obvious effect on the size of these aggregations. In high winds, the deer are distributed in small parties, sheltering in narrow gulleys and behind hillocks. In snow, large herds form in areas where the grazing is still clear. In summer, the activity of biting flies often appears to cause the animals to leave groups and to wander off singly.

Food dispersion, too, is important in group formation. When the deer are feeding on the herb-rich greens, they frequently collect in large, compact groups. But when they are feeding on the *Molinia*-dominated hill grazings, they are generally found in small groups within which individuals are widely separated. On the greens, the animals feed unselectively, moving steadily across the closely bitten turf. On the hill grazings, they are highly selective and move from clump to clump, choosing food with obvious care. In the latter case, the formation of large, compact groups might lead to mutual interference between feeding animals while on the greens this does not occur because food is more evenly distributed.

Such studies raise two important points concerning the relationship between the factors producing intra-specific variation in population density and those responsible for inter-specific differences. First, environmental factors may produce intra-specific differences in grouping which are not adaptive. For example, Kenji Yoshiba, working with Hanuman langurs in India, found that troop size and range were considerably larger in one area where population density was relatively low than in another where it was extremely high. Although it is possible that these differences were adaptive, they may have been no more than the effects of variation in population density.

Secondly, it is unsafe to assume that factors which produce variation in grouping at an intra-specific level are also important in producing similar differences between species. Marcel Hladik found that differences in group size and range size between Hanuman langurs and purple-faced leaf monkeys were associated with differences in diet. Hanuman langurs, living in larger troops in larger ranges than the leaf monkeys, fed to a greater extent on fruit and less on mature leaves, utilising food resources which were widely dispersed and seasonally variable. These differences in group and range size seemed likely to represent an adaptive response to the dietetic differences. In contrast, there was no evidence to suggest that the similar differences in grouping found by Yoshiba between populations of langurs were associated with ecological variation of this kind.

Thus all three methods of investigating the functions of differences in grouping patterns between species incorporate different problems. The present need is for studies, like Goss-Custard's, that combine several methods. When grouping patterns in a variety of animal species have been investigated in this way, it will be possible to make more definite statements about their functions.

14

Food and society in marmots and monkeys

'MONITOR'

29 August 1974

It is not always easy to see the connection between an animal's habitat and eating habits, and its social organisation.

It is now a commonplace to state that, in man, biological evolution has given way to social evolution. But whereas the environmental forces shaping biological evolution have been the subject of systematic study for decades, the factors that determine social organisation have only recently been subjected to serious analysis. Hence the rise of the new science of socioecology, which represents the attempt to show how the environment has influenced social groupings and behaviour patterns in a number of species other than man.

In this kind of enterprise, primates usually take star billing, because of the inevitable temptation to extrapolate to man – which makes it all the more amusing that, out of two articles appearing this week in *Science* and *Nature* respectively, it is the one on marmots, a genus of large North American rodent, and not the one on primates, that is both the more engagingly written and much the more anthropomorphic of the two.

David Barash, of the University of Washington, has spent the last seven years patiently watching three species of the genus *Marmota*, living in climates of varying clemency and predictability. The exigencies of climate and food supply have produced little difference in general appearance between the species. But Barash reports in *Science* (vol. 185, p. 415) that they have produced conspicuous differences in the animals' behaviour. The easier you have it, it seems, the more antisocial you are likely to be. Woodchucks, which live in low-lying fields and forests with long growing seasons in which to fatten up against a long hibernation, are solitary and aggressive. They confine their social contact with adults to the bare necessity of copulation and throw out their young promptly at a year, a very mechanistic view of life.

Olympic marmots, which inhabit higher ground farther north, where the growing season is shorter and the climate more capricious, are at the other extreme. They live in small colonies and acknowledge their neighbours each morning during the "visiting period" with a formal greeting. Young are allowed to stay with mother for at least two years, and sometimes three if a bad winter has resulted in high mortality in the colony that year.

Between the Olympic marmot and the woodchuck, Barash has recorded an entire range of yellow-bellied marmots, living in a graded series of intermediate environments. The proportion of polite to tolerant to aggressive behaviour of the yellow-bellies, and the length of time the mothers are prepared to wear their young, are broadly correlated with the harshness of the environment in which the animals have to survive.

By contrast with this neat and pretty story (French marmots may be a little out of line with the general theory but probably only because their habitat has recently been shifted by pressure from human pursuits), primatologists in Africa have been having a terrible time making any sense at all of the mere eating habits of colobus monkeys.

The difficulty, according to Tim Clutton-Brock (*Nature*, vol. 250, p. 539) is that the criteria for different dietary and social categories are not subtle enough. Both the black and white colobus, and its cousin the red colobus, for example, could be classified as tree-dwelling vegetarians. How, then, to explain that the black and white monkeys form little troupes of less than 10, which, when threatened, move noiselessly to cover, whereas the red colobus range in groups of 40 or so, which scatter, under attack, uttering loud shrieks and calls?

Clutton-Brock thinks he may have discovered an answer but only after going into the dietary preferences of the two species in nit-picking detail. The red colobus will eat only young shoots, flowers and fruit, but from a wide variety of plant species. Black and white colobus, on the other hand, will eat any part of just two plant species and no others. That means that a small area of dry forest is enough to feed a few black and white colobus, which do not need to winter in wet areas to find plants at the right stage of growth for shoots, fruit or flowers. A small area can be effectively defended by a small troupe. But the red colobus, with his predilection for seasonal delicacies, needs a wider and a wetter area in which to find them. Larger numbers may have the advantage not only of safety from predators, but also of increasing the chances of finding an acceptable plant at the right growth stage.

15

Apes, sex and societies

ALEXANDER H. HARCOURT
and KELLY J. STEWART
20 October 1977

Man's closest living relatives are the great apes – chimpanzees, gorillas and orang-utans. Despite their genetic affinities, chimpanzees and gorillas organise their social and sex lives in very different ways, so we need to be careful when looking to the apes for clues as to how human society evolved.

Soon after Darwin published his *Descent of Man* in 1871 there appeared, probably in *Punch*, a cartoon of a sobbing ape – half chimpanzee and half gorilla – pointing at Darwin and complaining bitterly that the great man was claiming to be one of his descendants. The artist's choice of a compound ape was very appropriate: without doubt, man's closet living relatives are the two African apes, the gorilla and chimpanzee. Fewer differences exist between some biochemicals of these apes and man than between similar molecules of other "sibling" species such as goats and sheep or horses and zebras. Chimpanzees and gorillas are closely related to each other too, so much so that some taxonomists want to put them in one genus, *Pan*, instead of the two they are in now, *Pan* and *Gorilla*.

Despite their similarities, chimps and gorillas behave very differently from each other and they organise their lives in markedly contrasting ways. In this article we want to compare the very different mating systems of chimpanzees and gorillas, and we will try to explain those differences in terms of their social organisation. When one compares species, it is best to view the many aspects of their behaviour as parts of a whole and not as separate entities. As a necessary background to our discussion of sexual behaviour and mating systems, we will first describe gorilla and chimpanzee social systems.

Chimpanzees live in what have been called "loose" communities in which, for much of the time, equal numbers of males and females move around separately. These days chimps live scattered in a belt across central Africa where they may inhabit environments that vary from relatively dense woodland to relatively open grassland. Male chimps travel, sometimes alone, sometimes with others, over the whole community territory which they defend communally against males from other communities. Not only do they threaten and attack trespassing outsiders, but it seems as if males even go on boundary patrols to check incursions.

One phenomenon that sets chimpanzee society apart from that of many other mammal species concerns the fate of the maturing males. When the young males of most mammals reach adolescence they leave the group in which they were born to spend their reproductive lives with unrelated females, thus avoiding any harmful effects of inbreeding. Years of work by Jane Goodall and her coworkers and students at the Gombe Stream National Park in Tanzania, and by Japanese investigators in the Mahali mountains, also in Tanzania, have confirmed that chimpanzees do not follow this pattern: males remain in the community in which they were born. Thus, males within groups are more closely related to each other than they are to males in other communities. This unusual arrangement has some interesting behavioural consequences which we will describe later.

Compared to the information available on male chimpanzees, we know far less about the females' social behaviour and ranging patterns. When adult chimpanzee females are not in oestrus they live in overlapping home ranges about half the size of the male's range. Many of these females are mothers with young, and they spend much of their time wandering with their offspring, or travelling with other mother–offspring pairs. Oestrous females, on the other hand, appear to prefer the company of males. Unlike males, the adolescent females usually do leave their home troop to join a neighbouring one.

It would be hard to imagine a more marked contrast to chimpanzee social organisation than that of the gorilla. Parties of adult chimpanzees very rarely stay together for more than a few days at a time, but gorillas live in stable groups that forage through the dense undergrowth and scattered trees of their leafy home ranges, also in central Africa. The home ranges of neighbouring groups may overlap at their edges.

Living in each group is at least one silverback (a fully mature

A dominant male or 'silverback' gorilla with two of his females

male), one-to-several adult females (some of whom will have been with the silverback for years) and their offspring.

As well as these differences, there are also some striking similarities between gorilla and chimpanzee social systems. Many of these similarities stem from the fact that they are the only two primate species we know in which it is the females rather than the males that habitually leave their natal group and transfer to other groups. This means that in chimp groups the males are generally related to each other whereas the females are not, a very unusual arrangement for primates. Similarly, in the troop of gorillas the females are unrelated to each other. Male gorillas present a somewhat different picture, simply because, unlike chimps, gorillas live in something approaching a harem: the dominant silverback male tolerates only one or two young males – probably his sons – who will take over leadership in the group when their father dies. The other young males leave the group and wander alone until they find mates for themselves.

If you watched a group of chimpanzees and a group of gorillas for a day, and if there was an oestrous female in each group, you would notice three differences between the species. One would be the extent of sexual dimorphism – the difference in shape of the sexes – in the two species. You need more than a momentary glimpse of an adult chimpanzee in order to be certain whether it was a male or a female: males are bigger than females, but not dramatically so and the two sexes are the same colour and general build. Adult gorilla males and females are absolutely unmistakable even at a brief glance, simply from their body size. The second difference you would notice would be the female's sexual swellings. The

A male chimp indicates his readiness to mate by waving branches and displaying his erect penis

chimpanzee female's is enormous, while the gorilla female's is so small it is usually imperceptible. The third difference is that you would almost certainly see the chimpanzee male's impressive court-ship display while the gorilla male has next to none.

There are many other contrasts too, but these three are the most obvious ones. Why these differences? To start, let us deal with the contrasts in size of the sexes.

Some adult male gorillas have a number of females in their group, while others wander alone and so have none. Unless these loners aquire females from neighbouring groups, and manage to keep their

In her turn, an adolescent female displays her pink and swollen genital region, which shows her to be in oestrus

females with them, they will have no chance of producing offspring. Not surprisingly, therefore, the males compete quite openly for females. They usually do this by chest beating, ground thumping and foliage slapping, and no one gets hurt. Sometimes, however, fierce fights do break out; the aftermath of these can be blood bespattered, flattened vegetation over many square yards, with the antagonists bearing wounds for days, sometimes weeks, afterwards. Dian Fossey has a photograph of a silverback's skull with another silverback's canine embedded in it near the eye socket that is ample proof of the potential violence of these contests. When fights

E

point: the winner is likely to be the larger animal. This is one reason between males are so severe, a large body size is an advantage up to a for the gorilla male's enormous size: he is the largest of the primates and twice as heavy as his females.

The other part of the explanation also concerns fighting ability but against predators, not competitors. In times of danger, the leading silverback of a group turns to face and, if necessary, fight predators. Again, a large male is probably better equipped to defend his females and offspring and so it is to the females' advantage to choose larger males.

The second selection pressure for large size does not apply to chimpanzee males, because in these animals females with offspring and males usually wander separately. At first sight there seems no reason why chimpanzee males within a community should not fight over females as frequently and fiercely as do gorilla males. Usually, however, they do not. Males, especially high ranking ones often try to prevent others from mating but their competition does not reach the level of gorilla males'. Why not? One answer is that, by fighting, a male can actually lose an oestrous female to other potential partners: there are always several chimpanzee males gathered round an oestrous female, and as one male fights with his competitor, a third might sneak in and take his chance. The other answer concerns the fact we mentioned before that chimpanzee males within a community are related to each other. As far as so-called inclusive reproductive fitness is concerned, it is not a good thing to fight with your relatives: a relative shares some of your genes, so if you harm the relative you decrease the chance of "your genes" being carried to the next generation. Gorilla males, who are usually unrelated, need have no such inhibitions.

From males' body size we now turn to the females' sexual swellings. First, one should realise that the whole point of the swelling seems to lie in its attractiveness to males: Jane Goodall describes a group of male chimpanzees who set off across a valley in pursuit of a female there exhibiting a large conspicuous pink backside: their intent was unmistakable! Female chimpanzees in oestrous have a number of males to choose from. Though an individual male probably knows most of her potential mates individually, she might not know them all that well, because she does not associate with them on a permanent basis. The swelling is not only a way of signalling to the males that a particular female is ready to mate, it is also a method by which females can compare the males and detect which mate they prefer. By attracting a selection of males around her, the female chimp can make her choice more easily than if she had to compare them one by one.

A gorilla female, by contrast, has already shown her preference for a mate: she transferred to the group that he leads. There is no advantage to her, when she is on heat, in attracting a number of males; all she has to do is walk up to her previously chosen male (who will rarely be more than 50 to 100 yards from her) and solicit. Thus the sexual swelling of the gorilla female is minute compared to the chimpanzee's: she has no special need to exhibit herself.

Chimpanzees' oestrous swelling persist for about 10 days, gorillas for only one to two days. Again, when a female chimp is fertile, it helps her to be attractive for longer to give the males time to gather and the female time to make her choice.

One supposes that it is advantageous to the female chimps to choose the most vigorous male to mate with. The males, competing with each other for the oestrous female's favour, show their readiness to mate by their energetic and flamboyant courtship displays. They demonstrate their vigour by erecting their hair and waving branches, and their readiness to mate by displaying an erect penis.

We were at Dian Fossey's Karisoke Research Centre in the Virunga volcanoes of Rwanda and Zaire for three years (our work being supported by the National Geographic Society, Leverhulme and Medical Research Council studentships and Mr and Mrs James Stewart), and in all that time we did not see any male gorilla behaving in a chimp-like fashion. For the leading silverback in a group, the problem of being chosen by an oestrous female from among other adult males does not exist. Once he has a female in his group he can be almost certain of being her only mate, and so he does not have to work so hard at attracting her. A soliciting female received next to no attention from the male until she was almost in his arms.

Performing courtship displays is not the only way that male chimpanzees compete for females. As we have said, they sometimes attempt to keep the females to themselves, and thus occasionally resort to aggression to do this. Caroline Tutin has found that most conceptions occur when a male and female travel in consortship away from all the other adults; nevertheless, an appreciable proportion of fertilisations do occur during times of promiscuous mating, when the males gathered round the female mount her in rapid succession. A male that could not take a female away on a consortship, nor prevent other males mating with her, could still manage to sire offspring if he could outcompete the other males during the "gang-bangs". One way he could do this would be to have more efficient fertilisation through more vigorous or more numerous sperm.

Mating efficiency is related to two other marked, though not immediately noticeable, differences between gorillas and chimpanzees. One is the difference in time they take over their copulations. A male chimpanzee under pressure from the others can mount, thrust, ejaculate and dismount, all within an average time of about seven seconds. Gorilla copulations are much more leisurely affairs: a pair usually join for more than one and a half minutes, and sometimes up to a quarter of an hour.

Mating is a leisurely affair for gorillas

The other contrast related to mating efficiency lies in the size of the animals' testes. A gorilla male is about three times as big as a male chimpanzee, yet his testes are only one sixth the size of the chimpanzee's. In fact, no primate yet measured has testes that are a smaller proportion of the bodyweight than a gorilla's. Many people have pondered on this difference between the two species, but only one person seems to have hit on the notion of relating it to the animals' mating efficiency and so to their contrasting mating systems. Professor Roger Short at Edinburgh thinks that the enormous size of the chimpanzee's testes could have resulted from selection for production of large amounts of sperm. The more spermatozoa a male produces, Short suggests, the higher the chance that one of *his* sperm will result in fertilisation. For the gorilla male, there has been little or no such selection pressure: the silverback is the only male likely to inject the sperm. To highlight the differences between the males of the two species, we can say that the chimpanzee has been selected for his ability to mate and the gorilla for his ability to fight.

Perhaps the main lesson to be learned from this short contrast of the mating systems of the gorilla and chimpanzee is that great care is needed in the sort of comparisons we make between animals and man. Here we have two very closely related species, yet their mating systems and social organisation could hardly be more different. We must not try to make direct analogies between only one of them and man. It is through comparative studies of *many* non-human species, not just gorillas and chimpanzees, that we may work out the principles governing their behavioural and morphological traits. These principles are the key to our understanding the biological bases and evolution of human behaviour.

Let us end by saying that a greater knowledge of our own species is not the sole aim to our studies. We strongly believe that in so far as a biologist's work benefits the wild animals themselves, his studies are valuable. Other species too have a right to live, and man will surely suffer if those other species disappear, even if only because some of the beauty and variety of life have been removed.

16

Gelada baboons and the battle of the sexes

ROBIN DUNBAR

6 July 1978

Male and female geladas both want to rear as many offspring as possible but the interests of the two sexes may often be in conflict. How is the conflict resolved?

The intrepid visitor to Ethiopia's remote northern plateau will sooner or later encounter one of the more exotic species of monkey, the gelada baboon (*Theropithecus gelada*). When I first came across a herd of these animals on the grasslands that border the mist-shrouded gorges, I had the unavoidable impression of organised chaos: several hundred animals moved in a dense front across the meadows feeding busily. So intent were the monkeys on what they were doing that it was almost impossible to detect any order or structure in the milling throng. Yet, hidden beneath the seeming confusion, I soon discovered, is one of the more complex animal societies. The characteristic features of gelada society, I suggest, owe their origins to the basic conflict of interest between the ideal reproductive strategies of each sex.

Through patient observation of the animals over several months, one notices that the seemingly amorphous herd actually consists of a number of harems. Each harem is typically made up of a single adult male, three to five females and their offspring, although they vary in number considerably. The smallest unit I observed consisted of just one male and one female, while in the largest there were 28 individuals, including at least a dozen adult females.

In strictly biological terms, each individual's main objective in life is to reproduce as often and as efficiently as possible. The more offspring it can leave behind, the greater genetic success it can chalk up. However, the strategies an individual can use to accomplish this objective differ. Conflicts of interest between individuals arise inevitably, both between the adult male and his females and among the females themselves. As a result, their behaviour is governed by a number of conflicting demands and loyalties. These conflicts are especially strong between the two sexes, because their respective contributions to the process of reproduction are necessarily different.

Females reproduce most efficiently when there is a lot of food about and when there is little tension between them and the rest of the herd. Reproduction imposes heavy energetic demands on most female mammals: they need to consume about 25 per cent more food than normal during pregnancy and about 50 per cent more during lactation in order to nourish growing fetuses or infants. At the same time, social tensions can disrupt their reproductive physiology causing a miscarriage. In gelada baboons, these hazards can work together. Social tension and harassment interrupt a female's feeding. Repeated interruptions during lactation, for example, can mean the difference between providing the infant with an adequate milk supply and not being able to do so.

Females partially solve this problem by staying near their own mothers and sisters rather than leaving the harem when they reach adolescence; male offspring, by contrast, do leave their mothers when they near maturity. So harems are usually made up of subunits of related females who are less likely to interfere with each other's reproduction, and who even protect one another from interference by unrelated monkeys. Indeed, females show the strongest tendencies to interact with their closest female relatives. Members of a harem maintain these and other social relationships by mutual grooming; as they groom they clean each other's fur, an activity they obviously find rewarding and relaxing.

The situation appears simple enough, except for the fact that the male also needs to maintain relationships with the females in his harem. Unlike the female, the male's reproductive success depends only on the number of females he has in his unit: the more females he has, the more offspring he can father. However, he must "own" a harem in order to reproduce at all. Young single males compete for harems, and harem owners are certain to be challenged from time to time. The result is usually a fight between the two males and, although the challenging male usually has an advantage in terms of fighting strength and skill, the outcome is actually decided by the females, who desert or remain loyal as a group.

Thus, the ultimate power lies in the hands of the females. A male's "control" over his harem depends on his maintaining strong

Grooming is an important part of gelada social life. Here a female grooms the much larger male

relationships with each female and he tries to groom with them all. (Indeed, during a fight, the harem owner divides his time between fighting off the young challenger and frantically grooming each of his females.) However, he must also groom them regularly on "ordinary" days. The penalty for not doing so is extreme: a neglected female may be willing to desert him in favour of the next challenging male. Because each female "demands" a minimum amount of grooming for her loyalty, a male may be kept very busy.

On the other hand, the costs and risks of transferring allegiance can be considerable for females as well and we do not expect them to undertake transfers lightly. A change of males upsets the established network of relationships and inevitably there is jockeying among the females for position after a fight. The stress created by this in-fighting may bring on a miscarriage; and sometimes small infants are even killed. Nevertheless, females in large units may actually benefit reproductively from a change.

For females, there is an ideal group size which balances the advantages and disadvantages of group life. As well as the advantages I have already mentioned, family groups also provide protection from predators. For this reason, daughters usually remain in their mother's unit, and the size of harems inevitably increases as further generations are born. However, as the number of adult females increases, they begin to interfere with each other's breeding success by competing for food and other resources and through the stress imposed by overcrowding. When the harem grows beyond the ideal size, females may, therefore, begin to find desertion to a smaller harem to their advantage, despite the short-term disruptions.

A gelada baby hitches a ride

On the other hand, females cannot simply transfer from one harem to another, because the resident females of the new harem are usually extremely hostile towards outsiders. For this reason, females transfer only to males who have few or no other females. When their own harem gets too big, it is usually taken over by several males, either in succession or simultaneously. These males divide the females between them and form several smaller units.

The male must therefore ensure that his females do not desert him, or, to put it another way, he must try to postpone the inevitable for as long as possible. Unlike some other baboon species, who forcibly herd "wandering females" back into the harem, gelada males can do this only by grooming each female regularly and for a sufficient amount of time.

In a large harem, the male finds this a difficult task. The amount of time he can devote to grooming is limited by other necessities, including feeding, and travelling between sleeping cliff and feeding areas. In a small group, the male is able to "cover" all his females easily by dividing his time equally among them. However, as his harem gets bigger, the total number he can give his attention to in a day becomes limited. Sooner or later, he finds himself in a dilemma.

If he continues to divide his time equally among the females, he grooms each one for less than the "loyalty minimum". On the other hand, if he concentrates on just those few whom he can groom properly, he is certain to lose the remainder.

Furthermore, females seem to be more anxious to groom with their favourite female partners than with their male. Indeed, the male often has to *insist* on grooming with them and after he initiates grooming, they usually terminate it to go back to their preferred partners. Females seem to value their relationships with the males less than their relationships with other females, perhaps because the male contributes only a small (although necessary) start toward her total reproductive effort.

Clearly, this network of conflicts fairly rapidly devolves into a vicious circle. As the number of females in a harem increases, the male must spend less time with each one. Released from the "burden" of grooming the male, females spend more time grooming their preferred partners and less time grooming other less preferred females. As a result, the harem as a whole becomes fragmented. Conflicts of interest between the female subunits increase, further weakening the bonds between them and the harem becomes increasingly susceptible to challenge from young males.

Losing a few females might not be of great consequence to the male, except for the fact that one female's desertion tends to precipitate an avalanche in the same direction, because a female's first loyalty is to her preferred grooming partner. If one partner deserts, the other partner (or partners) will go along with her. In this way, fights between males are in effect over the possession of all the females.

Cast unwillingly into Hobson's choice, the male is put into an even more serious predicament because he has one, and only one, opportunity to own a harem during his lifetime. Once he loses control over his unit, he has almost no chance to regain it or to capture a new one.

Although a harem owner's females remain loyal only as long as it is in their own interests, the actual timing of a desertion, of course, depends on the availability of a suitable bachelor male. If no male challenges the harem owner, he could conceivably remain in control indefinitely. As this is unlikely, however, the ultimate power to pick and choose lies with the females.

Of course, the animals themselves do not assess the pros and cons of their long-term biological strategies in the way humans may assess their career opportunities; if geladas make decisions at all,

Keeping his harem is a worrying business for the male – if he loses it, he will never get another chance

they make them in terms of immediate biological and psychological needs and propensities. These needs and propensities have been programmed genetically over generations by natural selection, so in many respects the analogy of conscious decision making is very useful in understanding these conflicts of interest. Indeed, such analogies are being used increasingly in research to uncover patterns of conflict in many species, patterns which may have had important influences on their social organisation.

17

Female chauvinist birds

DONALD JENNI

14 June 1979

The northern jacana is probably unique in the way it allocates sex roles. Females have up to four mates simultaneously and contribute little towards the raising of young; they fight each other for territories and males. Why do they do it?

Life is all about reproduction, but the two sexes often disagree on how best to solve the problem of leaving lots of offspring. The initial biological contributions of male and female are different, for while the male supplies only a minuscule sperm the female contributes the egg, a significantly larger investment. In birds the discrepancy is magnified, and by the time a fertilised egg is laid the female has put far more into it than the male. We might expect the male to desert the female and seek to breed with other females. But if an abandoned female cannot raise the offspring as successfully as a female who has help, the male will suffer if he leaves. So the usual pattern of reproduction in birds involves either close cooperation between the sexes in raising the young (monogamy), or else a female bringing up the brood more or less singlehandedly (often this is linked to polygamous males). Very occasionally, however, the system is upset, and we find cases of *females* abandoning their eggs to the sole care of a male; they may do this for a succession of males – sequential polyandry – or for a group of males at one time – simultaneous polyandry.

As far as I know, the northern jacana (*Jacana spinosa*) is the only bird that commonly practises simultaneous polyandry. A female may be "wife" to as many as four different males and sometimes copulates with all of them in less than an hour. Female jacanas are brazen hussies, liberated females, or fine examples of evolution in action, depending on your point of view.

The northern jacana, Jacana spinosa

Northern jacanas breed on floating aquatic plants on marshes, lakes and streams, from Costa Rica to northern Mexico and occasionally as far north as Texas. I have studied them intensively over a period of years at two locations in Costa Rica. One study site is in a dry tropical forest where the short rainy season limits breeding to a relatively brief period. At the second study site, which is located in a wet forest, the jacanas breed all year round; the dry season is short and not usually severe.

During the breeding season, jacanas hold territories on which they feed and breed, but at other times of the year they are highly social. Although males and females both perform nest-building movements, the females' building behaviour appears to be an empty pretence that has no effect on the nest, a collection of plant parts that the male puts randomly on top of one another on the floating vegetation. Some nests are so insignificant that one could not even find them except for the eggs, while others are more conspicuous and consist of a reasonable amount of vegetation. Some males go in for most impressive nest building but only after the eggs have been laid. When he returns to the nest, the male settles onto the highest part of the nest platform and rolls the eggs up between his wings and body. He then fills in the hollow left by the eggs, stretching forward to grab bits of vegetation with his bill and pulling them toward his chest. When he next comes back to the nest, he sidesteps around the eggs, compacting the vegetation, until he stops on the highest part and repeats the whole process.

Males alone incubate the eggs throughout the 28-day period.

Most females never even visit the nest site after they have laid their fourth and last egg, except to help the male chase away potential predators. A few females sometimes come on a visit, especially after a fight with a neighbour. They peer at the eggs and then may stand briefly over them. One particular female often lowered herself until she touched the eggs with her breast feathers; she would then straighten up and walk away. This was the closest I saw to "normal" female behaviour. I assume that a female visits the nest to reassure herself that her eggs are still there; jacanas often lose their eggs, accidentally and to predators, so the female who checks her eggs from time to time will be more successful in the struggle to reproduce.

When the eggs begin to hatch the male becomes extremely restless and vocal. Suddenly he will not tolerate the presence of, for example, purple gallinules (*Porphyrula martinica*), which prey on jacana eggs and chicks, not even in distant parts of the territory where he ignored them before. The male's voice alarm brings his female into the fray and she spends as much time as the male trying to keep predators away from the chicks. But the female never broods the chicks, nor does she accompany them when they feed, although the chicks sometimes attempt to follow her as she walks away from them. So what, apart from laying the eggs and seeing off competitors and predators, does the female contribute to raising a family?

Our data finally revealed that females perform one additional parental duty; they rest or preen within a few metres of the chicks 37 per cent of the time that the male is away from them. In contrast, when the male is with the chicks, the female is also near them for only 6 of every 100 minutes. Even when near the chicks, the females act totally indifferent to them, but immediately attack any potential predator that appears.

Although some females are monogamous, the majority of female jacanas in Costa Rica have two or more males. The average ranges from 1.7 to 2.5 males per female. In some places, where habitat quality is poor and males need large territories to provide enough food, or where territories are linear, as along streams, it becomes physically impossible for one female to defend two territories. The females are, effectively, monogamous.

Near the town of Turrialba, where the jacanas breed year-round, the number of males per female varies considerably from year to year as shown in Figure 1, and sometimes changes dramatically overnight. On the pond I looked at, each male defended his small territory against all other males and tried also to defend it against

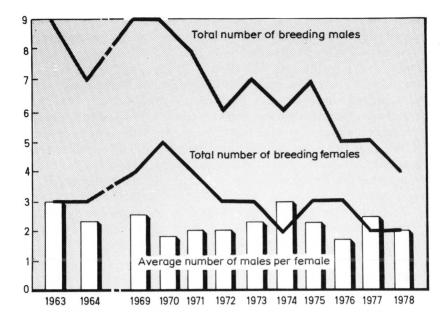

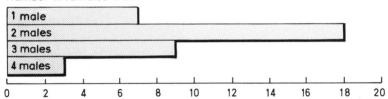

The number of males and females on the pond Jenni studied varied from year to year, but not always in step; so the number of males 'owned' by each female also varied. Most females had two males, though three did manage to breed with four males

intruding females. Males weigh, on average, only 91 grams while breeding females are almost twice as large, at an average 161 g and a male can do nothing to exclude a persistent female. Nothing, that is, except call his own female. It is up to the female to keep other females from intruding on the territories of each of her males. Females work very hard defending their males and lose possession much more often than the males. A female is usually displaced from the male she is defending by a neighbouring female, who then acquires the additional male and his territory. More often, though, a territorial female is evicted by a female who is not a breeding neighbour. In these cases, the interloper is inevitably larger than the resident bird. There are always a few non-breeding jacanas at the pond, just waiting for the opportunity to capture a male and start to breed.

In western Costa Rica, where the jacanas breed seasonally, all the suitable breeding habitat is occupied simultaneously at the beginning of the breeding season. As the rains continue, water levels rise and create additional suitable areas which are promptly occupied. Although the territories are large at the start of the season, each male becomes less persistent in defending distant parts of his territory as soon as he begins to incubate his first clutch. His reduced aggression makes it possible for a new male to take over part of his territory. This pattern of territory packing gives the impression that the female defends a large super-territory which is subdivided among her males. At Turrialba female territories sometimes expand or contract to include or exclude exactly one male's territory, and although females and males are replaced individually, new boundaries for both sexes generally follow pre-existing territory boundaries. Males defend their own territories against other males independently of female ownership and females defend males and their territories against other females.

Polyandry in the jacana is, quite fittingly, a female affair, but to achieve it the female behaves, to all intents and purposes, like the male of a polygynous species. She is bigger and more brightly coloured, and each male forms a pair bond by behaving subordinately towards whichever female succeeds in excluding other females from his territory. He has no chance to express a preference, as females work out who owns the males and their territories through aggressive interactions with one another. Surprisingly, there seem to be equal numbers of males and females, though obviously fewer females than males breed in any season. One might expect to get strange breeding systems, like polyandry, from strange sex ratios, but this is not the case for the jacana.

Although polyandry has been claimed by ornithologists and other natural history texts for decades, simultaneous polyandry had never been carefully documented prior to our work in 1972. There was, however, quite a lot of evidence for sequential polyandry – a female laying for a succession of males one after the other.

Charles Darwin heard tales – of females behaving as males and cooperating over laying their eggs – from Argentine gauchos; he offered the "simple explanation" that the female greater rhea (*Rhea americana*) breeds sequentially with more than one male during a single season. More than 130 years later we learned that Darwin's guess was right. Phalaropes are small shorebirds, like sandpipers, and all three species show sex-role reversal. The brightly coloured, aggressive females abandon their males after egg laying, and it was presumed that they did so in order to seek out additional males to breed with. After all, the reasoning went, sex-role reversal hardly makes any sense at all unless one takes full advantage of it! Some evidence suggested that the secretive tinamou, an ostrich-like bird that haunts the forests of South and Central America, has a system of mating similar to that proposed for the rhea by Darwin. There was also good evidence that the pheasant-tailed jacana (*Hydrophasianus chirurgus*) of eastern Asia was sequentially polyandrous. Having laid for one male, females consorted with and eventually copulated with and laid for a second male.

But why be polyandrous at all? For polyandry to be an advantage, the female must be relieved of at least a significant proportion of the chores associated with rearing the young. It is difficult to imagine how copulating with a variety of males could possibly increase a female's reproductive success if she still had to care for all the offspring. Mateship systems and parental care systems are inexorably related.

In some instances, the system of parental care determines the mating system. For example, if male and female both have to be around full time for the young to stand a chance of survival, monogamy will be obligatory. If, on the other hand, the male provides a smaller share of the total parental care than the female, he might be able successfully to help more than one female. In the circumstances, we might expect some kind of polygamy, with many females bonded to a single male. If the female cares for the eggs and young on her own, the male could be either monogamous, as in most ducks, or polygynous, as in most game birds. In those rare cases where the male performs all parental care, the female could be monogamous or, if she can produce multiple clutches – no mean feat given the resources that go into an egg – she could be polyandrous. If she mates with only one male at a time, sequentially, paternity never seems to be in doubt and the male has nothing to lose if he is able to rear the chicks. One could consider such "polyandrous" matings as sequential monogamy.

The important point is that the relationship between the parental system and the mating system is usually permissive rather than obligatory; although sex-role reversal does not demand polyandry, polyandry cannot exist without it. Naturalists often fail to distinguish between the mating system and the parental system, which has generated a great deal of confusion because a species with reversed sexual behaviour is not necessarily polyandrous. A missing mother phalarope does not have to be out courting another male.

In 1972, the year in which we first described simultaneous polyandry in the northern jacana, sequential polyandry was carefully documented in three other species of shorebirds, including, at long last, one of the phalaropes. In that same year an entirely different sort of polyandry was described in the Tasmanian native hen (*Tribonyx mortierii*). Although a few of the native hens are monogamous, most females form bonds with two or more males, usually brothers. These birds live together in expanded family groups that share territory, nest, and clutch. Although both brothers copulate with the female, one copulates more often than the other. There is no sex-role reversal, and this system of fraternal polyandry resembles the system of "helpers at the nest" as practised by the Florida scrub jay. The brothers help each other, and the strategy is successful; polyandrous native hen females raise more young than do monogamous females.

How did polyandry evolve? Several shorebirds lay two lots of eggs (double-clutching) and this is a good starting point for the evolution of polyandry. In its simplest form a female produces a clutch which she leaves in the care of a male and then lays a second clutch which she cares for herself. The female's second clutch could be fertilised by her first mate or by a second male who already has a clutch that was laid by another female. Or, instead of caring for the second clutch all by herself, the female could pair with a second male and cooperate with him to raise her second clutch. Clearly, the capacity of a female to produce second clutches rapidly is as important a precondition to polyandry as reversal of the sex roles. Other than this, we do not really understand why some birds have become polyandrous.

Polyandry is reported in a few human cultures and a consideration of human polyandry should throw light on the same phenomenon in birds. Unfortunately the term polyandry has been applied as carelessly to human societies as to avian societies. There has been a tendency to identify as polyandrous any society in which women regularly form multiple sexual alliances. In many of these societies the men also form multiple sexual alliances and these expanded groups of mates are really neither polyandrous nor polygynous; you could almost call them polygynandrous. In some, the alliance bonds are simultaneous, in others the alliances are sequential and some human societies are, like the rhea, simultaneously polygynous but sequentially polyandrous.

All the human mating systems that closely resemble avian poly-andry are based on fraternal polyandry. The primary function of fraternal polyandry in humans appears to be to keep family wealth within the family. However, there are two very different socio-economic conditions under which fraternal polyandry has arisen. In some cultures the brothers take a single wife and their children are raised as a single family around the mother in order to avoid subdividing the family estate and wealth among a number of brothers and their families. This system occurs among the wealthy class of serfs in some areas of Tibet. Alternatively, one wife may have many husbands in particulary harsh settings, where the burden of supporting a wife and raising children is so great that one man is unable to do so, and brothers marry a single wife. In these cultures, in which brothers share all their property, it is consistent for them also to share sexual access to a single wife. This system resembles the fraternal polyandry found in the Tasmanian hen, and both systems have been given the descriptive name "wife sharing".

In fraternal polyandry, human and avian alike, there is no reversal of the sex roles, and wife sharing is never accompanied by the kind of matriarchy found in the northern jacana. Fraternal polyandry is based on cooperation between brothers, who share genes, and is clearly a male strategy. Males do better by aiding their brother, than by breeding themselves, perhaps without help. In the northern jacana the males share nothing with one another except a common territory boundary where they occasionally agree to fight. The males are subordinate to the females and show complete sex-role reversal. The jacana's simultaneous polyandry is a female strategy and is maintained by the females. Save that they do not lay eggs, I would be tempted to call the males "female".

18

The song of the great tit says "keep out"

JOHN KREBS
3 June 1976

Great tits sing in order to defend territories against intruders. But their songs are too complex to be simple warning signals. Why?

What does bird song mean? Although poets through the ages have extolled the beauty of the nightingale or blackbird in full voice, birds sing not for aesthetic reasons, but because song aids the individual in the cut-and-thrust struggle for survival and reproduction. Birds produce a great variety of sounds, ranging from simple croaks and squeaks to the more complicated melodies that are referred to as song. True song is restricted to the song birds (oscines) and in contrast to most simple call notes, such as those given at the approach of a predator, it is largely produced by males.

In temperate regions song is seasonal, that is, more common before and during the nesting season than at other times of year. The bird that I work on, the great tit, has a rather simple song compared with many species. It consists of a repeated phrase of about three notes which is usually written as "teecher teecher teecher" in field guides and it is sung, nearly always, by the male sitting on a prominent perch, from January till mid-May, when the young hatch out.

In most song birds there is a rough seasonal correlation between the time they begin singing and their reproductive activities such as territory establishment and pairing. This correlation has led most people to assume that song is concerned either with proclaiming territorial ownership or with attracting a mate, or both, the relative importance of the two varying from species to species. Although these ideas seem reasonable, there is remarkably little direct evidence for either of the supposed functions of song. It is well known for many species that a territorial male will respond vigor-

Great tits are fiercely territorial. Here one displays aggressively to warn off an intruder

ously by attack or displaying at a loudspeaker broadcasting his own species song within his territory. This shows that a territory owner can recognise the song of his own species and identify it with a potential rival but it does not tell us whether song acts, as everyone has assumed, as a deterrent to prevent intruders from trespassing into a territory. In my research, I set out to test experimentally whether song acts in this way as a "keep out" signal.

The great tit is primarily a bird of woodland, although it is also common in farmland, gardens and orchards, wherever there are trees to provide suitable nest holes and foraging places. In the autumn and early spring young birds and some adults compete for territories in the best breeding habitat, which in my study area near Oxford is mixed deciduous woodland. Unsuccessful competitors usually end up by settling in hedges and gardens of the surrounding suburban and agricultural land, where breeding success is apparently lower than in the wood. I had previously found that if I removed territorial residents from woodland, new pairs appeared rapidly and took over the empty spaces; some of the new birds came from territories outside the wood, and others seemed to be non-territorial "floaters". The new arrivals often appeared within a few hours, so they must have regularly monitored the woodland to check for empty places and I thought it likely that one of the cues they used in this monitoring was the song of territorial residents.

To test whether this is what really happens, I set up the following experiment. I plotted the territories of all the pairs of a six hectare copse of woodland on Wytham Estate near Oxford (see p. 61), and then on one morning in February I captured and removed all eight pairs. I now "occupied" three of the territories with loudspeakers broadcasting the song of the former residents, while the other territories acted as controls – two with a control sound and three with no loudspeakers. If song deters intruders, the territories which were occupied with loudspeakers broadcasting song should be invaded more slowly than the controls. I repeated the experiment again about a month later, reversing the positions of the treatments. Each experimental territory was occupied with a tape recorder playing a continuous loop of tape through an amplifier linked by a multiway switch to four loudspeakers in different parts of the territory. The loudspeakers approximately mimicked the singing pattern of a real bird: each speaker was active in turn for two minutes to give eight minutes of song, which was followed by 52 minutes of silence. This hourly cycle of singing and silence continued in each territory from dawn till dusk, starting more or less immediately after I had captured the resident pairs.

The results of the two experiments were similar. On both occasions the two control areas were invaded before the experimental territories (Figures 1 and 2). Within about one day, new birds were singing and showing territorial activity in the control areas. The experimental area was also colonised eventually, but not until after about two and a half days, so the loudspeakers were effective in keeping out intruders in the short term. In other words, it seems that intruders listen to song as a means of assessing the density of resident birds. Obviously the loudspeakers did not simulate the presence of a real bird very accurately, so that the intruders soon realised that the "occupied" territories were empty.

If song carries the simple message "keep out", why is it so complex? In most species of song bird, each male has a repertoire of different versions of his territorial song. In birds such as the great tit and chaffinch, the repertoire is small, each male having less than half a dozen different song types, while at the other extreme song thrushes, blackbirds and robins may have repertoires of several

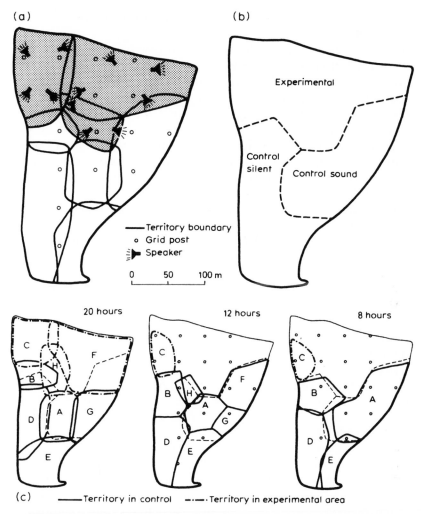

Figure 1 (a) *The positions of the territory boundaries in the wood before the start of the first experiment. The approximate positions of the loudspeakers are shown in the experimental territories. (b) The wood was divided into experimental, control sound and control silent territories. In the control sound territories, the loudspeaker broadcasted a two-note phrase similar to the song of a great tit but played on a tin whistle. (c) The progress of reoccupation of the wood by new birds. The number of hours refers to the number of daylight hours since the start of the experiment (about nine and a half hours per day). The letters in each territory refer to the sequence in which new birds arrived, and the dotted line shows the boundaries between the three treatment areas*

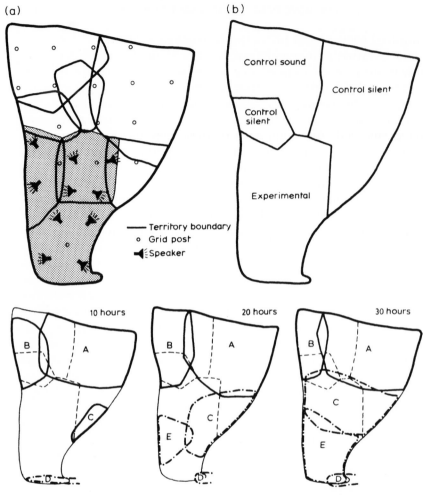

(a)

- —— Territory boundary
- o Grid post
- ◄ᶓ Speaker

(b)

Control sound

Control silent

Control silent

Experimental

10 hours

B | A

C

D

20 hours

B | A

C

E

D

30 hours

B | A

C

E

D

—— Territory in control area —·—Territory in experimental area

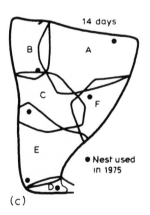

14 days

B | A

C

F

E

D

• Nest used in 1975

(c)

Figure 2 (a) and (b) The positions of the territory boundaries in the second experiment, with the approximate positions of loud-speakers in the experimental territories. (c) As in the first experiment, the control areas were occupied before the experimental territories. In this experiment only four pairs replaced the eight I had removed, although two weeks later an extra bird settled. A sixth bird had part of its territory in the wood hundred songs. Most often, the different song types are used in the same context; a singing male may work through his repertoire in a few hours or even minutes, so it is hard to imagine that the different songs contain different messages. Even a small song repertoire seems to be redundant. Many people have speculated on the problem of the significance of song repertoires and four main ideas have emerged. Although each of these hypotheses may apply to some cases, it is not likely that one hypothesis will explain the evolution of all song repertoires.

One suggestion is that repertoires provide variability which allows individuals to recognise one another by song. I consider this to be rather unlikely. Individual recognition has to be based on features of song which are constant within an individual but vary between birds, and if each bird has many different songs, then the chances of any feature being constant must be small, so that repertoires may actually *decrease* the chance of individual recognition. A second idea is that a repertoire enables a territory holder to direct its "keep out" signal to a particular intruder. Several workers have noticed that when song is played through a loudspeaker, the territorial resident male will sing back at the speaker using a song in his repertoire which matches the playback. One interpretation of this is that matching is a way of saying "I am responding to you" instead of "keep out". To test this, we would have to find out whether there is any difference in the reaction of the intruder on occasions when the resident matches and when he does not. In general, this explanation is more likely to apply to very large song repertoires than to small ones since only by having a very large repertoire could a territory holder be reasonably sure of being able to match any intruder that happens to come along.

A third hypothesis is that song repertoires are analogous to peacocks' tails, that they have evolved by sexual selection. If females are attracted to males with large song repertoires and males gain an advantage as a result, sexual selection could favour extreme development of elaborate songs. The problem with this is that many of the species with large repertoires are monogamous, and sexual selection is not so likely to be important in monogamous species because a male can gain less by attracting a single female than he could by attracting many females in a polygamous or promiscuous species. In the great tit, I could find no evidence that males with larger repertoires succeeded in attracting females that laid larger clutches or bred earlier in the season, two possible advantages a male could get from sexual selection in a monogamous species.

Finally, varied repertoires might act as an anti-habituation device. If a male broadcasts his "keep out" message again and again in the same way, perhaps the listening intruders will eventually come to ignore it in the way that all animals tend to wane in their response to a repetitive stimulus, through the process of habituation. I have found that territorial male great tits do indeed habituate more rapidly to repetitive playback of a single song than to playback of a repertoire. The difficulty with the habituation argument is the assumption that habituation to songs is an inevitable constraint on the auditory system.

It is much more likely that habituation is an adaptive pheno-menon moulded by natural selection and we know from work on many animals that habituation to biologically important stimuli is very slow, so why should intruders habituate to the songs of terri-torial residents? One suggestion is that habituation is part of the mechanism by which intruders assess the density of birds in an area. When a bird is trying to set up a territory in early spring, it presum-ably "shops around" looking for a good breeding habitat where the density of birds that have already settled is not too high.

The loudspeaker "occupation" experiments show that the intruders use song as a cue for assessing density, and habituation may be a mechanism by which an intruder chooses to settle in a low-density area: the fewer song types it hears, the fewer birds are likely to be present and the more likely it is that the new bird will be able to establish a territory. So by habituating more rapidly to a smaller number of song types the new settler is able to choose a good place to settle. The advantage to the territory holder of singing a repertoire of songs is that he, in effect, causes the potential new arrivals to overestimate the density of singing birds and so try to settle elsewhere. This is, of course, only speculation but it should be possible to test the idea experimentally by comparing the response of new settlers to territories occupied with loudspeakers playing repertoires of song types and territories with single song types.

It seems, therefore, that although bird song is one of the most widely studied and appreciated communication systems in nature, the problem of why song is so complex is still an unsolved mystery.

19

Song is a serenade for the warblers

CLIVE CATCHPOLE

2 April 1981

Unlike the aggressive great tit, the sedge warbler sings its intricate songs in order to attract a mate. Other species may have songs that fulfil both functions.

"Naturalists are much divided with respect to the object of singing in birds." So wrote Charles Darwin in *The Descent of Man and Selection in Relation to Sex*, and the remark is as true today as it was then, with "biologists" substituted for "naturalists". True, modern equipment has enabled biologists to analyse bird songs into component syllables, and has revealed a variety and complexity of sounds far greater than the human listener suspects. But such detailed analysis merely compounds the problem and to the perennial question, "Why do birds sing?" we must now add "Why are their songs so elaborate?"

Biologists traditionally divide the vocalisations of birds into short, simple *calls*, produced by either sex at any time, and longer, more complicated *songs* produced mainly by males in the breeding season. The simpler calls pose no great problems: they are made in particular contexts and, as the calls have an immediate effect upon the behaviour of other birds of the same species, biologists have found it less difficult to interpret their meaning. Thus, calls can tell other birds about the species of the caller, the caller's sex, personal identity, location and even his or her motivational state.

But if simple calls — albeit extended into a vocabulary — can convey so much information, why have some birds developed the far more elaborate vocalisation of song? Why, among the groups of passerine (perching) birds known as oscines, or true songbirds, has evolution favoured the production of such complex sounds, much of which seems to be redundant?

Biologists have advanced three main explanations to account for the high level of redundancy in bird songs. The first is the non-selectionist view, that the seemingly functionless elaboration is some kind of avian music, and that birds compose variations for their own pleasure, or even to entertain ornithologists and musicians. Certainly, no one would deny that bird songs are extremely beautiful and do please the human ear, whether that ear belongs to a poet or a scientist. But beauty and function are not mutually exclusive, especially when the task in hand is to select a mate. Then again, the act of singing exacts a cost: it can make an otherwise cryptic (inconspicuous) species extremely conspicuous to predators. As songbirds characteristically sing in a daily rhythm (see Figure 1), their singing represents an enormous investment in both energy and time. The biologist is left in no doubt that natural selection would not encourage such behaviour, unless on balance the benefits more than outweighed the considerable cost and risk.

The second explanation is that the more elaborate signals transmit information that cannot be encoded in simple calls, and

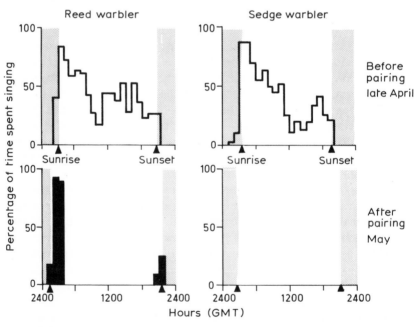

Figure 1 *Warblers sing at a regular time each day. After mating, the reed warbler sings far less than before, and the sedge warbler stops singing altogether*

that it is just a matter of time and technology before we crack the code and reveal some sophisticated communication system akin to human speech. Despite a great deal of high technology and more detailed analyses in recent years, this idea has led to a dead end. Bird songs appear to be no more or less like human language than other animal noises.

We are left with the final possibility, that although bird songs serve much the same functions as the signals of other animals, these more elaborate forms do transfer information more efficiently and effectively in particular situations.

The message contained in the song, and its meaning, may well depend upon whether the recipient is male or female. Females, as potential mates, may be attracted to the song, while males may be repelled from the songster's territory. Biologists have long argued about which of these functions prevails. Darwin suggested that the song's main purpose was to charm females but ever since the renowned British naturalist Eliot Howard published *Territory in Bird Life* in 1920, most ornithologists and biologists have embraced the idea that song serves primarily as a territorial proclamation (see p. 132).

Indeed, there are many reasons why the territorial explanation of bird song holds sway. The simplest reason is that when it comes to fighting and mating, birds are much like human beings. Territorial disputes are noisy, public confrontations which are easily observed, whereas sexual behaviour is a more quiet, private affair, well hidden from public view. Biologists have found it relatively easy to observe and record interactions between males during territorial disputes – the border skirmishes of blackbirds in gardens are a conspicuous example – but in many species the more subtle process of female choice has often been overlooked or ignored.

As their name suggests, warblers (family Sylviidae) are renowned among ornithologists for the beauty and complexity of their songs. For several years now, I have been studying the structure, functions and possible evolution of song in one particular group of European warblers – those of the genus *Acrocephalus*. This group is collectively known as the reed warblers, and there are six species in Europe: reed and great reed warbler, the sedge warbler, and the moustached, aquatic and marsh warblers.

Acrocephalus species are small, cryptic, and live in dense marshland vegetation. Because they cannot readily see each other, there is clearly a premium on vocal rather than visual communication. The first clue that song may be more important in sexual attraction came

Sedge warbler, Acrocephalus schoenobaenus

from studies on their seasonal and diurnal (daily) rhythms of song production. *Acrocephalus* warblers are migrants, and arrive at the breeding areas in Europe during spring. Immediately after he aquires a suitable territory, a male *Acrocephalus* starts to sing throughout most of the day and night. The rhythm of his song (Figure 2) is characteristic of unmated males among all species. Females arrive several days later and, in most species, including the reed warbler, the males sing far less after the pairs have formed. The male sedge warbler is particularly extreme in that he never sings again once paired (Figure 1).

Playback experiments, in which loudspeakers and recorded songs are used to simulate a rival male, confirm that male reed warblers still sing back after they are mated, as part of their territorial defence. Male sedge warblers also defend their territory after mating but they use visual threat displays and real fighting, not song. This suggests that the main function of song in both these species is sexual attraction but that some species continue to sing in defence of their territories.

These functional differences in song can often be explained by subtle but important differences in the ecology and behaviour of individual species. For example, reed warblers breed in tiny, close-packed territories in dense reed beds, where continual acoustic defence is clearly advantageous. Sedge warblers defend larger, more scattered territories in more open habitats outside the reed beds, where visual signalling is extremely effective.

Detailed sonagraphic analysis of song structure also suggests that sexual selection has played an important role in their evolution. Simple calls consist of one syllable which appears as a single, continuous mark on the sonagram trace. In contrast, the songs of typical songbirds such as chaffinches or great tits consist of a stereotyped sequence of several syllables and last between one and five seconds. When compared with these, the songs of *Acrocephalus* species, such as the sedge warbler, are astonishingly long, complex and varied. The sedge warbler song shown (Figure 2) is actually quite short, and some songs contain several hundred syllables and last well over a minute. Each individual has a repertoire of up to 50 different types of syllables, but uses only between five and 10 in any one song. The possible permutations are astronomical and, instead of transmitting a stereotyped signal, a sedge warbler literally composes a continuous stream of constantly changing variations. The only visual parallel that bears comparison with such extravagance is the peacock's tail and Darwin suggested that such exotic devices must result from female choice. That is, the evolutionary force behind the elaboration of tail or song is that of sexual selection; putting it crudely, it is the flamboyant males whose genes are carried into future generations.

The hypothesis that the song of the sedge warbler is some kind of acoustic peacock's tail is one that can be tested. It should be possible to find out whether discriminating females are still exerting sexual selection pressure in wild populations. If they do, then females will select males with more complicated songs, in preference to those with simple songs. But to test this prediction, we first need an objective measurement of song complexity.

One obvious measure of complexity is the size of the individual's repertoire of syllables: the larger the repertoire, the more numerous and complicated are the songs he can construct. At first sight the task of identifying and counting the syllables from recordings is daunting. But in practice, although a warbler may construct many

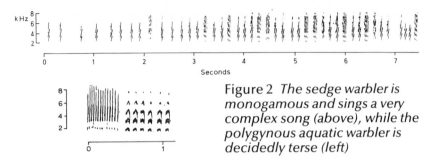

Figure 2 *The sedge warbler is monogamous and sings a very complex song (above), while the polygynous aquatic warbler is decidedly terse (left)*

hundreds of different songs, he tends to cycle through his entire repertoire of syllable types in about the first 10 songs. It is therefore relatively easy to arrive at a reliable estimate of syllable repertoire size from a sample of recorded songs from each individual.

I recorded the songs of each male sedge warbler soon after he arrived at the breeding area in spring and then followed closely the arrival of the females a few days later. I knew when the birds formed pairs, because the males immediately stopped singing, and this was later confirmed by finding their nests. I then plotted the date of pairing against the estimated size of the males' repertoire, and found a strong inverse correlation (Figure 3). It does indeed seem that newly arriving females are attracted first to males in the population with the most complicated songs.

Not all species of *Acrocephalus* warbler produce such complicated songs as the sedge warbler, but some species have even more complicated songs. By comparing song complexity within the same genus, I felt it should be possible to glean some further clues as to the possible evolution of their songs.

One major difference between species, and one which has obvious significance in the context of sexual selection, is in the type of mating system. Most birds are monogamous, as both parents are needed to bring food to a maximum number of dependent nestlings. But in certain rich habitats such as marshland, where insects may be superabundant, there is no need for both parents to help feed the young. In these circumstances another mating strategy can evolve – polygyny. Polygyny – literally, "many females" – implies that each male may mate with several females, but not necessarily remain with any one of them. Polygyny is possible when food is plentiful because even if a male does desert a female to seek another, the first female may well be able to feed and raise the young successfully herself. Polygyny has evolved twice in *Acrocephalus* species: in the aquatic warbler and the great reed warbler.

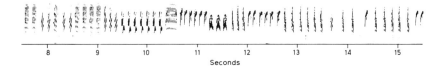

8	9	10	11	12	13	14	15

Seconds

If it really is sexual selection that has promoted the evolution of elaborate bird song then it might be thought that polygynous males will have evolved more elaborate songs due to the consequent increase in sexual selection pressure. To test this prediction, I applied the same technique for estimating the size of the repertoire of syllables to the songs of all six European warblers. But the results turned out to be a real surprise. The two polygynous species, the aquatic warbler and the great reed warbler, had a repertoire range of only 10 to 20 syllables and they sang relatively short, simple songs – as illustrated by the aquatic warbler in Figure 2. It was the monogamous species – reed, sedge, marsh and moustached warblers – that had by far the most complicated songs. Their total repertoires were within the range 35 to 100 syllables, and instead of singing short songs they produce long continuous bouts of song, each containing hundreds or even thousands of syllables (see Table 1).

Although at first puzzling, the apparent paradox can be explained by examining the different ecology and behaviour of polygynous and monogamous species, and by understanding the different ways in which sexual selection pressure might operate upon song structure in each case. It seems to me that in a polygynous species, a

Figure 3 *Warblers with the most complex song were soonest mated*

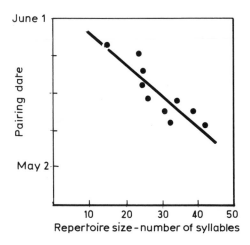

F

Table 1 Repertoire Ranges, Song Lengths and Mating Systems of Six European Acrocephalus *Warbler Species*

Species	Repertoire range: number of syllables	Mean length of song in seconds	Mating system
Marsh warbler (A. palustris)	80–100	continuous	monogamous
Reed warbler (A. scirpaceus)	70–90	continuous	monogamous
Moustached warbler (A. melanopogon)	60–80	continuous	monogamous
Sedge warbler (A. schoenobaenus)	35–55	19.49	monogamous
Great reed warbler (A. arundinaceus)	10–20	3.20	polygynous
Aquatic warbler (A. paludicola)	10–20	2.18	polygynous

female is less concerned with the quality of the male himself, as she is quite likely to be deserted and left to raise the young alone. To maximise her own reproductive success she must be more concerned with the territory and the food that it must provide for her young. Hence the female should not select her mate directly but indirectly – through the quality of the territory that he has obtained. Polygynous males defend large territories and may attract several females to nest within them. Male song has, therefore, evolved primarily within the context of competition for territories and *intra*sexual selection pressure between males. Mate attraction may still be an important, but secondary, function of song. The relatively short, simple songs of polygynous warblers are much more like those of more typical territorial perching birds, whose song has an obvious territorial function. Shorter, simpler songs may be more suited to territorial functions for a number of reasons. For example, they are more easily learned by neighbours who tend to keep their distance throughout the breeding season.

In a monogamous species, the territories are quite small and the birds collect most of the food from outside them by making long, arduous flights to neutral foraging areas. To maximise her reproductive success, the female is less concerned with the quality of the territory and much more directly concerned with the quality of the male. She should select a good-quality male who will help her in the important task of feeding the young. There are many cues she might use to judge the quality of a male, but in a species which uses song as the main method of attraction then the quality of the song itself could be an important early indicator. Once good-quality males start being more successful by using more complicated songs, then the runaway process of *inter* sexual selection leading to extreme elaboration could be set in motion. This is just what appears to have happened in the four monogamous species (reed, sedge, marsh and moustached) and direct evidence from the sedge warbler now supports this view.

Earlier studies on bird song have tended to emphasise the territorial function to the exclusion of mate attraction, and have even suggested a dichotomy between the two. The sexual selection theory presented in this article recognises the central importance of mate attraction as the ultimate goal of all such behaviour. It also resolves the false functional dichotomy by placing territorial behaviour within the same general framework as mate attraction. Territorial songs are then seen as the result of indirect intrasexual selection pressure; sexual attraction songs are seen to result from direct intersexual selection.

A sexual selection theory of bird song is unlikely to provide such a convincing explanation when applied to all bird groups, and one reason is that the two main functions of mate attraction and male repulsion are not mutually exclusive. It is highly likely that evolutionary economy and compromise have combined to produce song structures in some species that reflect both these basic requirements and in differing amounts. Teasing apart their relative importance will give the evolutionary biologist enough problems for countless sleepless nights. For him the infinite variety of the dawn chorus is a doubly cruel awakening.

PART FOUR

Theories and Models: Natural History Transcended

Collecting data is only half of research – good science means developing theories to make sense of the data. The theory of evolution gave ethologists a unifying principle around which to organise their observations. But it needed further development before it was powerful enough to predict most animal behaviour (p. 151). The wealth of data that had accumulated by the 1970s included instances that seemed hard to fit in. Why do the worker castes of social insects slave their lives away in the service of their queen and never reproduce themselves? Darwin's theory stipulated that every individual should be out to maximise its own fitness – it did not admit of behaviour that occurred for the good of the species. Recent theoretical advances have solved this problem by supposing that natural selection acts on the genes themselves rather than the individual that bears them. So it is just as profitable to help a closely related individual to reproduce as it is to reproduce oneself (p. 161).

Theories in biology are now being stated in increasingly precise mathematical terms. It is becoming possible to draw up balance sheets of the costs and benefits of almost any form of behaviour. The costs are energy expended and access to food or mates lost; the benefits are energy gained and opportunities to reproduce. Under the right circumstances, these costs and benefits can be given exact mathematical values, or at least be expressed in terms of probabilities. Reducing everything to numbers means that computers can be brought into play to cope with the more tedious calculations. But it is still up to the ethologist to base computer programs on sound theory.

One of the most fruitful theoretical advances has been in the application of game theory to animal fights (p. 171). Few fights between animals are real trials of physical strength and in many cases no one gets hurt. So what determines who wins? Game theory suggests that how hard an animal is prepared to fight depends on how much it has to lose. Out-and-out aggression entails the risk of injury or even death, so the stakes must be very high before a real fight is worth it. It is often better for the members of a population to exploit some inequality between contestants – such as age or experience – to determine who wins. The inequality can be as ir-relevant to fighting ability as whether the contestants are fighting "at home" or "away" (p. 182). Ethologists are now finding uses for game theory in considering behaviour without an aggressive element, such as cooperation.

The balance-sheet approach is also useful in assessing how animals make their day-to-day decisions about, for example, collecting food. The world is full of uncertainties: the most fruitful foraging patch today may be the least fruitful tomorrow. Because of this, it is not necessarily the best use of an animal's time to eat everything in sight as rapidly as possible. As Tim Beardsley explains (p. 190), animals seem to cope with uncertainty rather well, apparently "knowing" when it might be worth taking a calculated risk.

20

How animal behaviour became a science

DAVID McFARLAND
18 November 1976

As *New Scientist* passed its 20th birthday, students of animal behaviour were consolidating the wealth of data they had amassed and beginning to develop theories to account for it all.

The *New Scientist* was born 20 years ago, and its first words on the subject of animal behaviour were as follows:

Some animal behaviourists have got into the habit of wrapping their observations up in such tortuous terms that it is to be wondered whether even their fellow animal behaviourists know what they are about. The purpose, presumably, is to gain objectivity by avoiding phrases that have any connotations in terms of human conduct; but if they continue in this way they are obviously going to use words eventually that have no connotations at all. . . . This wouldn't matter so much if animal behaviour was a cosy little scientific sideline, but it isn't. The chances are that it will throw light on a hundred dark corners of evolution.

Whether animal behaviour studies have enlightened our view of evolution, I will discuss shortly. It is certainly true that 20 years ago students of animal behaviour were exercised about objectivity, as can be seen from the polemics of the time. However, science develops as a result of natural selection, the survival of ideas and methods being dependent upon selective forces operating in a future environment, which is largely under the control of the younger generation of scientists. Thus the growth of animal behaviour as a scientific discipline can be seen as an evolution of standards of evidence and of argument about the behaviour of animals.

The problem of objectivity in animal behaviour studies begins with the reporting of behavioural observations. In the early days these were largely anecdotal. A major textbook published 20 years ago read like this:

I am indebted to Dr Lorenz for some interesting unpublished observations of this kind. He tells me that a jackdaw will remember the hiding place of food for more than a week, and a raven for probably much longer; but that these appear exceptional amongst tame birds . . . and he tells me that a raven dislikes being observed when hiding food. If observed, it will take the food out again and scold the observer, and it will very soon learn to hide the food out of human reach. Lorenz finds magpies and crows intermediate between ravens and jackdaws in this respect.

This type of evidence is not as easy to evaluate as it might appear. It might seem that if an animal was observed to perform a particular act, one observation is surely enough to confirm that the individual is capable of the act. I am reminded of an occasion when an undergraduate burst into my room in a state of great excitement. He reported that he had seen a pigeon feeding from a slice of bread. When some other pigeons approached, it quickly pecked a hole in the bread, put its head through the hole, and flew off with the bread around its neck.

One thing that animal behaviourists have learned over the past 20 years is that there are many activities that animals can be observed to do that we have little hope of explaining in the near future. This change in attitude is well illustrated by a few words from a review paper published in 1970: "The animal behaves as though it thinks that consumption of the substance made it sick." The point is that a distinctively flavoured substance was ingested a number of hours before the animal was made sick by the experimenter. The animal immediately acquired a long-lasting aversion to the substance. To the young ethological puritan of 20 years ago the statement would seem to be begging the question of how we know how a rat would behave if it did in fact think in such a manner. In the present-day context the statement appears merely as a neat summary of the situation, and the authors are not concerned with the question of whether the animal consciously associates the two events. They know that research into such questions is not likely to pay off, and they have a good idea as to what lines of investigation would be fruitful. As a result of this general attitude, we now know, six years later, a great deal about learning in situations where there are long delays between behaviour and its consequences (see p. 22).

Many factors have contributed to the growth in confidence, pragmatism, and sophistication of animal behaviour research over the past 20 years. One is the enormous growth of factual knowledge, the product of careful, unhurried and dogged observation and experiment. It appeared, a few years ago, that this had been

Konrad Lorenz with his devoted greylag geese

carried too far, that phenomena had become more important than explanation, and that the subject was becoming structureless. However, it now appears that theory is catching up and that we are about to enter a new phase of interpretation of animal behaviour.

The revolt from the philosophical issues of the past has been accompanied by a growing expertise in the evaluation of data, and in the development of theory. The evaluation of data involves not only the proper use of statistics, but also the design of experiments and the proper use of controls. In animal behaviour research, this can be especially difficult, because animals are so complex. A fascinating study of the way in which this aspect of the evaluation of evidence can be important, is provided by a recent review by James Gould of the honey bee dance-language controversy. Gould notes that the dance-language hypothesis of Karl von Frisch was so successful in gaining acceptance that it never had to become very precise. When it was attacked on both theoretical and experimental grounds by Wenner and his colleagues, many years of controversy ensued, in which both sides repeatedly failed to employ adequate controls in their experimental attempts to justify their position. There is more than one way by which recruitment of workers is accomplished by bees returning from foraging. A bee can advertise a food location with site-specific odour information but it can also use a symbolic dance language.

Von Frisch had never denied that bees used odours, but only when Wenner and his colleagues challenged the evidence did it become apparent that the scientific world had, for many years, accepted the dance theory, even though the evidence was inadequate. However, whereas Wenner has expressed the widespread attitude that an animal is a simple system, and can behave only in the single simplest way, Gould recognises that the earlier tendency to anthropomorphise animals has been over corrected, and that too much scepticism may become a blinding bias against yet undiscovered abilities of animals.

Animals are much more complex than the behavioural scientists of 20 years ago were prepared to allow. For example, it was long assumed that an animal's various motivational tendencies, to feed, to court, and so on, competed with each other for behavioural expression. Thus if an animal chose to feed, it would be deduced that feeding was its dominant motivational tendency at the time. Recent experiments show that this is by no means always the case and it is now recognised that animals can interpolate subdominant activities for short periods of time, which remain under the control of the dominant motivation. Thus, a courting stickleback may take time off to check its nest, and the time is budgeted for by the dominant courtship control system. The recognition that competition is not the only way in which animals deploy their behavioural options has led to a whole new method of analysing an animal's motivational priorities, and has made it possible to formulate models in which animals can perform complex time-budgeting strategies, in which they can pursue more than one goal simultaneously.

The models of 20 years ago were generally very crude and simplistic, and suffered from a lack of professionalism and of understanding of basic explanatory philosophy. For example, 20 years ago a fashionable idea was that some kind of motivational energy, analogous to physical energy, was responsible for driving behaviour, and for the temporal characteristics of behaviour. The analogy breaks down because the behavioural energy was regarded as a causal entity, which energy can never be. Behavioural analogies to physical energy can be rigorously formulated, but these relate to functional aspects of behaviour, rather than to causal mechanisms.

An exception to this early primitiveness was the elegant application of control systems ideas, as practised in Germany. Unfortunately, it was many years before the philosophy of this approach was thoroughly digested elsewhere. A particular stumbling block was the question of whether behaviour should ultimately be explained in physiological terms. Control theory offers a type of explanation that can be applied to the study of systems in general, be they physical, biological or economic. Such explanation is independent of the hardware of the system concerned, and is a truly behavioural type of explanation. In relation to animal behaviour, this type of explanation is abstract in the sense of being independent of the particular physiological processes responsible for behaviour. To those raised on a strictly reductionist philosophy, this is a hard lesson.

Recently, behavioural scientists have begun to realise that standards of evidence apply, not only to conclusions drawn from analysis and evaluation of data, but also to the way behaviour is explained. They recognise that a model should be more than an elegant redescription of the data. It should also embody an element of hypothesis, which introduces parsimony into the explanation, and from which testable predictions can be derived. Moreover, a good model should be robust in the sense that predictions follow from the model almost regardless of its quantitative details. Robust models are not open to the charge that they can be manipulated to predict anything one wishes. Although a successful model is one whose predictions fail to be falsified, the standard of evidence attained is much higher if the predictions are *a priori* more vulnerable to falsification. Such models are said to have high "predictive information value".

In the process of growing up, animal behaviour has become a more unified and coherent subject, partly due to deliberate efforts to synthesise the fields of ethology and comparative psychology, and partly because these two disciplines learned from each other. It is fair to ask what science in general has learned from the study of animal behaviour. Over the past 20 years, behaviour has shed its mysticism and is treated like any other aspect of animal life, subject to natural selection and open to rational thought and experimental investigation. The contribution of animal behaviour to medicine was recognised, in 1973, by the award of Nobel Prizes to von Frisch, Konrad Lorenz and Niko Tinbergen. It would be disingenuous to pretend that this did not come as a shock to the iconoclastic younger generation of behavioural scientists. But they had failed to realise how the influence of ethology had subtly pervaded other areas of science. For example, the two most important recent advances in our understanding of evolution, namely the development of the concept of inclusive fitness by William Hamilton, and of evolutionarily stable strategies by John Maynard Smith, have been

Niko Tinbergen, one of the founders of the modern science of ethology

nurtured in an ethological climate. Although neither would call himself an ethologist, the significance of their discoveries has emerged from our knowledge of altruistic behaviour, on the one hand, and of ritualised aggressive behaviour on the other.

It cannot be said that the growth of animal behaviour as a scientific discipline has been without blemish. There has been too much naked aping of the behaviour of the scientific guru, able to diagnose the ailments of human society, and to dispense the remedies. Whiffs of metaphysical gas are still occasionally let off, and linger in the air, like the exhaust fumes of some too fast roadster. The rest of us experience a mild olfactory irritation, and move on, secure in the knowledge that such drivers generally take the wrong turning.

What of the future? Edward Wilson has prophesied that ethology and comparative psychology "are destined to be cannibalised by neurophysiology and sensory physiology from one end and sociobiology and behavioural ecology from the other". This is wrong for two fundamental reasons. In the first place, Wilson is incorrect in classifying ethology as "the naturalistic study of whole patterns of animal behaviour". The essence of ethology is that it combines causal and functional explanations. Whereas psychology deals exclusively with questions of causation, and sociobiology offers purely functional explanations, ethology alone attempts to show how form and function are related. For this reason ethology will always remain an essential lynchpin in the scientific understanding of animal behaviour. It may well be that the numbers of ethologists will decline, because the future growing points of the subject, such as ontogeny, motivational integration and cognition, require talents of a different order than those normally turned out of biology schools.

The second point is that Wilson envisages that "whole patterns of animal behaviour will inevitably be explained within the framework, first, of integrative neurophysiology, which classifies and reconstructs their circuitry, and, second, of sensory physiology, which seeks to characterise the cellular transducers at the molecular level". This exhibits a touching faith and a certain naivety. Few behavioural scientists seriously believe that such explanation is possible. The concepts required for a proper understanding of behaviour will not be explained in neurophysiological terms, because they relate to inherently hardware-independent phenomena. In 20 years' time, it is to be hoped that readers of the *New Scientist* will be able to judge for themselves.

21

In search of the rules of behaviour

'MONITOR'

10 May 1978

New theories of animal behaviour pinpoint the key features of an animal's environment that have the power to predict how that animal will behave.

Look at what sort of food an animal eats and how the goodies are distributed around the countryside, and, with a few more relevant data thrown in, it is possible to predict whether the male of the species is bigger than the female and what their sex life is like. This – put *very* crudely – is what Tim Clutton-Brock and Paul Harvey claim is now feasible (almost, anyway), given the increasing sophistication of theories on animal behaviour and the growing weight of research data with which to test them (*Nature*, vol. 273, p. 191).

Explanations of animal behaviours, particularly differences between species, have, say Clutton-Brock and Harvey, "been bedevilled by the absurd and the fantastic". Worse even than fantasies, though, is the misapplication of theory.

Although the business of gazing into the evolutionary past in order to guess the behaviour of animals long since extinct may seem a particularly intractable exercise, the task with modern extant creatures is not that much easier. There are many interacting variables, some of which one can identify, while others remain tantalisingly obscure.

In their *Nature* paper Clutton-Brock and Harvey look at four aspects of mammalian biology – population density, group size, breeding system and sexual dimorphism – in an attempt to demonstrate associations between differences between species and environmental variables. These four variables are interrelated, and Clutton-Brock and his colleague outline one way in which they are linked: mammal species live in groups or are thinly dispersed, according to the distribution, density and quality of the food they eat; whether animals are gregarious or unsociable is then very important in determining what type of breeding system (mono- gamous or polygamous) is possible; in its turn, the style of sex life will affect the animal's bodies (coloration, size, presence of antlers etc).

A key factor in now being able more accurately to explain (or predict) animal behaviour is, say the authors, the combination of energetic and genetic theory. Energetics involves two considera- tions. The first is the fuelling needs of individual animals – the bigger you are the more food you need but the increased fuel demands do not rise proportionately to the body weight because of reduced heat loss. The second is the type of packets the food comes in – animals that eat fruit and flowers will not be able to achieve as high a population density (or, rather, total biomass) as others that feed on the more abundant parts of plants such as stems and leaves. As plant foods are always more abundant than potential prey animals, carni- vores are more thinly distributed around the countryside than are the various species of herbivore.

Because large animals eat more food than small ones they are usually more thinly distributed over the ground; they often travel farther to get their daily diet too (they have a large home range). But the distribution of the food packets is very influential in determining the size of a species' home range. Animals that eat food that is "clumped" (a fruit-laden tree, for example), widely dispersed, or unpredictable in abundance, are generally forced to range far and wide. But if the daily diet is evenly dispersed and predictable, a species may thrive in a small home range. Furthermore, if the evenly distributed food is sufficiently abundant the species may find it profitable to defend a feeding territory.

The genetic element in the argument comes from recent develop- ment in ideas about evolutionary "strategy" of individuals. First, an individual can be said to be successful in biological terms if its genetic stock is maximised in the future. This means not only having as many offspring as is compatible with their survival, but also exercising some concern for brothers and sisters because they are genetically related. This obviously has some effect on the interaction between related individuals living in the same group.

The second genetic aspect may be termed the battle of the sexes. More specifically, the facts of biological life are that, in mammals at least, one parent (the mother) invests more in the offspring than does the father — she not only carries the offspring during gestation but suckles them for a time after birth too. For the female there is a very real practical limit to the number of offspring she can hope to produce. For the father there is, theoretically at least, no limit. In terms of social organisation, this explains why, in mammals, polygyny is common whereas polyandry is extremely rare.

A male can maintain a harem only when the distribution of food resources allows the females to be in close company with each other, either in a permanent home range or as a relatively mobile group. In other words, the male must be able to defend either his female entourage or a feeding territory where they live.

Monogamy, a rare form of social arrangement in mammals, occurs either where the male cannot command resources sufficient for more than one female, or if the upbringing of the offspring demands the care of more than one parent. Generally, though, if ecological circumstances are favourable for polygamy, the males take advantage of it.

The last part of this whole equation focuses on how sex life affects physical appearance. The point is that in a species whose social organisation involves harems (one-male groups, to be more technical) there is a good deal of competition to be the top male. The prize for the winner is great — sexual access to a large number of females; the loser has nothing and his genetic future is bleak. This competition often manifests itself in an exaggeration of physical form: male silverback gorillas are twice as big as their mates, red deer stags have enormous antlers, and the peacock has its absurd tail, all of which are designed to impress potential rival males, and possibly the disputed females too.

Sexual dimorphism is, of course, not as simple as that. You can not be outrageously big if you spend your life up a tree, for example. To complicate matters, females have a biological "incentive" to be large as it helps in collecting sufficient food in order to sustain their ever hungry infants.

There are exceptions to the "rules" but, undoubtedly, rules *are* emerging. With the accumulation of more data against which to test them, they will become even more well defined.

22

Genes take care of their own

JEREMY CHERFAS
4 January 1979

The concept of "inclusive fitness" helps to account for apparently altruistic acts among animals. The more closely related they are, the more willing animals should be to help one another. But, recently, examples of cooperation between unrelated individuals have come to light that need an alternative explanation.

Biology is replete with examples of phenomena that seem hard to explain in terms of the survival of the fittest. Take the selfless sacrifice of the worker honey bee in the service of her queen. The worker is sterile and cannot reproduce, yet she toils to nurture the offspring of another bee. She is even willing to die in defence of the hive. How can one account for this altruistic behaviour that so obviously goes against the fundamental principle of life – reproduction is everything? One might speculate from the comfort of one's armchair that the worker makes her sacrifice on the altar of the species. She does what she does because it enables honey bees to survive.

But think what would happen if a mutant bee, one who cheated the system, were to arise. Such a bee might be resistant to the substances the queen secretes that keep her workers sterile and might herself reproduce, allowing others to care for her young. A cheat bee might also hang back when the hive is attacked, secure in the knowledge that other suckers will defend her. This cowardly, reproductive bee would have cowardly reproductive offspring and in no time the selfless workers would be outnumbered by cheats. Who then would care for and defend the hive? What would happen to the species if no bees were prepared to raise the young and the hives were always open to attack? They would no longer be social honey bees, though some lines might survive as solitary bees.

A "good of the species" argument is intuitively very appealing, but nevertheless it is clearly impossible. New developments in biology have led to real understanding of altruism in general, not just as it applies to bees, and along the way have suggested solutions to many other knotty problems.

The keystone of Darwin's evolution by natural selection was the idea of fitness. In every generation there are those who are better adapted to deal with the problems posed by life. If the environment is becoming colder, animals with a thick coat will be better adapted than animals with a thin coat. More of the thick-coated animals will survive and breed, and if thickness of coat is determined by genes, the proportion of thick-coated animals in the population goes up. Fitness, crudely, is the ability to leave offspring that will in turn survive to breed and any change that increases fitness will be selected.

Worker bees are clearly not fit, for they leave no offspring at all. How could evolution ever produce an animal that is sterile? Inclusive fitness, an idea that owes much to J. B. S. Haldane but was fully developed by Bill Hamilton, provides the answer. Where Darwinian

Lionesses in the same pride are closely related, and help to rear one another's cubs

fitness concentrates on the individual, inclusive fitness concentrates on the individual's extended family. To understand inclusive fitness we need to know a little about reproduction and genetics.

The so-called genetic blueprint is contained in the chromosomes which, with the exception of the sex chromosomes, come in pairs. Man has 23 pairs of chromosomes, the fruit fly only eight. Genes exist in alternative forms, called alleles, and an individual normally has two alleles for any given gene, one on each chromosome of a pair. (This is not, as it happens, the case with bees, where the females have only one chromosome of each pair while the males have both, and although this changes the mathematics of what follows it does not change the conclusions.) Thus, to take one of Gregor Mendel's famous examples, there is a gene for seed-coat texture in the pea and it has two alleles, smooth and wrinkled. A pea can have two smooth alleles, or two wrinkled alleles, or one of each. In reproduction the offspring receives one chromosome of each pair from each parent, and it is a matter of chance as to which of a parent's two alleles for each gene the offspring gets. Thus, from the point of any single parental allele there is a one in two chance that it will be passed on to the offspring. This probability is the same for all the genes in all the chromosomes, so that the relatedness (Hamilton's term) of an offspring to either one of its parents is one in two, or a half.

All offspring of the same pair of parents are related to each parent by one half, but how are they related to each other? Well, looking again at just one gene for the sake of simplicity, there is a one in two chance that a given allele will be passed on to offspring A, and a similar chance that the same allele will be given to offspring B, so the chance that it is present in both A and B is one in four. If A and B share one parent they will be related to each other by one quarter, as there is a one in four chance that they both have the same copy of any particular gene. If they share both parents their relatedness is one half, because they are related by one quarter twice. This sort of quantification can be carried out for all sorts of relationships. Grandchildren, for example, are related to each of their grandparents by one quarter, and cousins by one eighth.

All these sums are fine and dandy, but how do they help us to understand evolution? The relatedness of two individuals is the probability that they share a particular allele and the purpose, if you will, of each allele is to maximise the number of copies of it that survive. Two full siblings are likely between them to contain a copy of any one of your alleles, so it matters not *to the allele* whether you or your two sibs survive to breed. It is this view of relatedness that led to Haldane's celebrated pronouncement, made in a pub after some hasty scribbles on the back of an envelope, that he would lay down his life for two of his brothers or eight of his cousins.

Inclusive fitness, then, is the total fitness of all your relations, devalued appropriately by their relatedness to you. If you help two of your sibs to have 10 offspring each it is, in terms of inclusive fitness, exactly the same as having 10 offspring of your own. The concept is not restricted to "relations" as we normally think of them, but the mathematics get harder as the blood ties get weaker, and the best examples of inclusive fitness in action do refer to quite close relations. In such a case the selection process is called kin selection, to distinguish it from Darwinian individual selection.

The worker bee is helping to rear her own sisters. In forgoing reproduction and assisting her mother she maximises her inclusive fitness, and ensures the survival of many more copies of her genes than she could on her own. Inclusive fitness helps us to understand the evolution of sterile worker castes in insects, where the family ties are rather clear. It also provides a powerful analytical tool for taking apart other societies, especially those that include a lot of behaviour that looks to be altruistic.

A pride of lions provides a perfect example of a society that apparently runs on cooperation and altruism. The females form the stable core of the pride. They hunt together, and as a result are able to capture large enough animals sufficiently often to sustain the whole pride, including sick lions who make no contribution to getting food. While some females are out hunting others stay behind with the cubs, suckling all and sundry rather than solely their own offspring. These babysitters will also feed from the kill, as will the males, who don't do much of anything. From time to time male rule over the pride is challenged by young bloods from other prides, and occasionally the challenge is successful. The old males are driven out, often to die, and the new males take over. When this happens the incoming males often kill the cubs in the pride.

The lion pride defies comprehension in simple evolutionary terms, but thanks to the painstaking work of Brian Bertram in the Serengeti National Park we now have some understanding of how the pride works. The problem in conveying this understanding lies in knowing where to break into the life cycle of the lion. I will start with the arrival of new males in a pride.

Behind the successful takeover of a pride will be a combination of factors. The more challengers there are, the more likely it is that they will succeed. Success is also more likely if the resident males are old and few in number. The new lions often kill off the cubs. This is understandable, as the cubs are the offspring of the previous incumbents, and their continued survival will only delay the start of the new males' reproductive effort. Lions hold a pride only once; it is their sole chance to breed and any behaviour that allows them more breeding during their short tenure will be selected. What is not so understandable is why the females don't put up much of a fight when their young are being massacred. After all, the cubs represent an investment for the females, and though they could not fight the males directly, one can imagine a different kind of pride in which the lionesses would help the lions to withstand a challenge if there were young cubs in the pride. The reason the females do not behave like this will become apparent.

New lions in a pride are also able to get rid of cubs as yet unborn; after a takeover those lionesses that are pregnant often abort. This is akin to the Bruce effect that was first demonstrated in laboratory mice, and is probably due to chemicals in the urine of the new males.

The death of the cubs, born and unborn, means that the lionesses quickly become fertile and sexually receptive. Moreover, they all tend to come into oestrus at more or less the same time. They copulate with the new males and eventually give birth to their cubs, again more or less synchronously. We now have several cubs and several lactating mothers in the pride. A lioness will give milk to cubs other than her own, so the more females there are with milk the more likely it is that a cub will be able to feed while its mother is away on a hunt. For this and other reasons, cubs born out of sync often do not last very long. Females can benefit from the death of their previous cubs if it means the present cubs have a better chance to survive.

When the lionesses make a kill, and it is they who make most of the kills, all members of the pride are able to eat, as long as times are good and kills frequent and large. When times are not so good the lions come along and bully their way to the food. If there is any left, the lionesses get it, and only then do the cubs get a look in. This is tough on the cubs, who often starve, but it ensures that the lionesses get enough food to sustain the next hunt. Besides, when supplies improve, it is easier to produce a cub, with its short period of gestation, than a full grown animal capable of killing the improved supplies.

If times are good the cubs will grow and mature. What happens to them next depends on whether they are male or female. Females will be accepted into the pride if there is room, but otherwise will be driven out. If they are very lucky they may find an unoccupied space in which to set up a new pride, but by and large these outcast females never breed. Males are invariably driven out of the pride. A cohort born together will wander about looking for a pride to take over, closing the circle.

Recall that a larger group of males stands a better chance of getting a pride and keeping it. They will father more cubs as a result. Recall too that males who kill cubs engender synchrony of births in the females, and that cubs born in synchrony are more likely to survive. The more cubs that survive, the larger will be the group of males driven out to find a new pride. They will be more successful and so cub-killing will spread through the population.

Communal suckling would spread for similar reasons. Perhaps at first a lioness's daughters would stay with her in the pride, so that in suckling other cubs the lioness would be caring for her grand-children. Cubs survive better in larger groups, so as she suckles other cubs the lioness ensures that her own offspring grow up with companions. Her sons will hold on to prides longer and her daughters will be more successful because they too will suckle other cubs. Infanticide on the part of incoming males synchronises the females, and makes it easier for communal suckling to occur. Put the two tendencies – males who kill cubs that are not their own and females who suckle cubs that are not their own – together, and you have a cooperative social system that will spread through a popula-tion and, once established, maintain itself. If the females that started the creche were related to each other, as sisters, or mother and daughters, they would be helping their relatives, and the system would spread even more quickly.

Bertram watched four prides for several years, and built up a composite picture of the typical pride. From such a pride he calcu-lated that the males were more or less half brothers (related by about $1/4$) while the females were approximately full cousins (related by $1/8$.)

Brian Bertram's work shows how the various threads of lion life can be unravelled, and inclusive fitness provides a way to see how each thread contributes to the fabric. By concentrating on the separate bits of behaviour, each of which appears at first sight to decrease the individual fitness of the animal concerned, we see how they fit together in a way that in fact increases the likelihood that they will survive into the next generation.

Not all societies are lion prides, but all are founded on similar principles. Among monkeys, there are many species in which males take possession of a group of females, but only a few in which the new males kill the infants. A wolf pack contains related females, but only one of them breeds. Hunting dogs cooperate to bring down large prey, but even when times are hard it is the pups that have priority at the kill. Slowly the similarities and differences between various societies are beginning to make sense, and they do so in the light of inclusive fitness.

The secret of our new understanding of behaviour, especially social behaviour, is to forget about the individual and think about the gene. Copies of an individual's genes are likely to be present not only in its own offspring but also in its close relatives and their offspring. Any gene that strengthens the tendency to help relatives will assist in its own spread and survival. The tendency to help relatives seems to be at the bottom of all societies. And at the bottom of the greatest revolution in biology since 1859 is Bill Hamilton's notion of inclusive fitness.

23

The secret of the lions' cooperative society

'MONITOR'

28 April 1983

Coalitions between male lions sometimes include unrelated individuals. This poses a problem for the theory of kin selection but might be explained in terms of game theory.

Male lions are not the solitary, ferocious beasts that inhabit the jungles of Hollywood. Whereas females cooperate in hunting and in rearing cubs, male lions cooperate with one another in forming coalitions which try to gain and keep possession of groups of females. Despite their weaponry, males within a coalition rarely fight over oestrous females.

Brian Bertram, now Curator of Mammals at the London Zoo, first suggested explanations for this surprising harmony. A male who stands by and allows another companion member of the coalition to mate with a receptive female is not really being altruistic because that companion is probably a relative. As William Hamilton, now at the University of Michigan, pointed out in the 1960s, a male can perpetuate its genes in the next generation, not only by fathering offspring himself but also by assisting the reproductive efforts of near relations who share many of his genes. Evolution can thus favour apparently "altruistic" behaviour which increases the reproductive output of relatives, if this behaviour really does increase the individual's inclusive fitness – the total representation of its genes in the next generation. This evolutionary driving force was dubbed "kin selection", to distinguish it from the additional (and inevitable) shaping of an individual's own reproductive efforts by natural selection.

In the case of lions, not only was a rival male likely to be a close relative, but the costs of physically competing with him were high and the benefits low. Costs were high because injuries to a male can

easily be crippling or fatal, whereas serious injuries to his companion can leave him without an essential partner in combating
rival coalitions. Benefits were low because only a minute proportion
of copulations result in a reared cub in the next generation (*Journal
of Zoology*, vol. 177, p. 463).

Bertram, working with David Bygott and Jeannette Hanby of the
University of Cambridge, showed that males actually did increase
their lifetime fitness by cooperating with one another (*Nature*,
vol. 282, p. 839). They followed the fates of individually known
lions living in the Serengeti and Ngorongoro Crater in Tanzania
over several years. Their data showed that the larger the coalition,
the greater was the fitness of each male within it, despite the fact that
he had more companions with whom he had to share paternity. The
effect resulted mainly from the fact that larger coalitions could
maintain tenure of a pride for longer. Because, they argued, the
males in coalitions were almost always close relatives, kin selection
enhanced further the benefits of cooperation.

But is kinship really an important driving force behind cooperation in male lions? Craig Packer and Anne Pusey of the University of
Chicago have continued to follow the life and loves of these
Tanzanian lions (*Nature*, vol. 296, p. 740). Now, it seems, non-
relatives are not so uncommon in coalitions; 5 out of 12 (42 per
cent) of the coalitions they observed contained non-relatives –
considerably (but not significantly) higher than the 10 per cent
(2 out of 21) recorded in Bygott, Bertram and Hanby's study. Packer
and Pusey suspect that the likelihood of finding non-relatives in a
coalition probably varies with the average age of its members, since
young related males may die and be replaced by singleton males
from other prides. The coalitions studied by Packer and Pusey often
contained older males.

Packer and Pusey observed oestrous females attended by rival
males, sometimes relatives and sometimes not. They showed that
disputes between the males were somewhat more common than had
previously been reported. They also noticed that males within a
coalition did not discriminate between kin and non-kin: they did
not compete more intensely with non-relatives. This observation,
they believe, throws doubt on the importance of kin selection. The
researchers suggest that game theory (see p. 171), rather than kin
selection, can account for the general absence of fighting over
females within a coalition. Game theory suggests that contests over
oestrous females will be settled "conventionally", by the recognition of asymmetries such as "owner versus rival". The costs of

fighting, which often leads to serious wounds or blinding, would outweigh the benefits of mating with a particular female in the long term, they argue. Males compete mainly, say Packer and Pusey, by trying to anticipate oestrus in a female and become the first to consort with her.

Bertram argues that game theory and kin selection should not be seen as alternative explanations (*Nature*, vol. 302, p. 356). "The role of kin selection is that, in competition between related males, it makes the costs higher and the benefits lower". Nor is kin selection the only force promoting cooperation in lions but it can amplify the initial advantage of being in a group, he maintains. The lack of discrimination between kin and non-kin within a coalition may indicate, Bertram suggests, that their behaviour is only roughly "tuned" – good enough, perhaps, in a situation where the great majority of interactions are among relatives.

24

Playing games is a serious business
ANTHONY ARAK
2 February 1984

Animals use rules to settle contests, and the winner may be determined with little regard to which is the better fighter. The application of game theory to animal behaviour, pioneered by John Maynard Smith, is now being extended to forms of behaviour other than fights. It may help to explain cooperation between unrelated individuals.

To the layman, the expression "nature red in tooth and claw" probably conjures up an image of animals engaged in a bloodthirsty struggle, all against all. Konrad Lorenz, author of *On Aggression*, commented on a film he had seen in which a Bengal tiger was wrestling with a python, and immediately afterwards the python with a crocodile: "What advantage would one of these animals gain from exterminating the other? Neither of them interferes with the other's vital interests."

Lorenz's statement reflects the fact that, except in films or on our television screens, animals rarely display aggression against members of other species unless, of course, one of them wants to make a meal of the other. But aggression between members of the same species is a different matter. Because members of the same species share similar habits and have similar needs, we can expect them to compete in earnest for those resources essential for survival and reproduction. A naive prediction of evolutionary theory might be that animals would benefit by murdering their rivals at every opportunity to obtain access to scarce resources, such as food, territories and mates.

This prediction, however, raises some difficulties. Conflict and fighting between members of a species very rarely take the form of outright aggression. On the contrary, ethologists cannot help but be impressed by the restrained and gentlemanly nature of animal fighting. In the spectacular fights of fence lizards, for example, a male grabs his opponent by the head and shakes it vigorously. After a short bout of wrestling, however, he lets go and allows his opponent to take a turn. Even animals armed with dangerous "weapons" such as sharp teeth and horns seem to fight to Queensbury's rules. Bighorn rams charge at one another and bang their heads together with a resounding thump. If the aim is to defeat the opponent, surely it would be better to wait until his back was turned and then butt him in the rear!

In some instances, animal combat is strictly a non-contact sport. Sheep and deer often use their horns and antlers for ritualised displays; many carnivores raise their hackles and bare their teeth without resort to physical combat. The function of many animal weapons seems to be to *prevent* serious battle by establishing hierarchies of dominance that individuals can easily recognise and obey. Fights involving injury seem only to occur when failure to obtain a resource may mean that the animal starves to death or fails to breed at all.

Early ethologists were often puzzled by the variable nature of animal aggression and could offer no satisfactory answer to the question, What is aggression good for? On the one hand, they argued that ritualised aggression in which blows are never exchanged would be good for the species because it would prevent injury and save lives. On the other hand they suggested that fighting to the death, when it does occur, is a good thing because it ensures the survival of only the fittest members of the species!

Two developments in evolutionary thinking have helped to resolve the question of why animals do not usually go all out for the kill during fights. Biologists now realise that the crucial question is not whether an aspect of animal behaviour is "good for the species" but whether it is "good for the genes" that cause individuals to indulge in it. Genes specify "behavioural strategies" and, therefore, it is the differential survival of genes that will determine whether or not a particular strategy is maintained in a population (see p. 161).

The second development came with the application of game theory to questions about evolution. This approach, pioneered over the last decade by Professor John Maynard Smith at the University of Sussex, has provided great insights into the nature of animal aggression. The essence of Maynard Smith's approach was to find the "evolutionarily stable strategy" for fighting behaviour, or the ESS for short. This is the strategy which, once evolved in a population, would be *stable* against invasion by other possible strategies.

The importance of the concept of stability for understanding the evolution of animal behaviour cannot be overstressed, for even a strategy that seems good for individual survival and reproduction will not be maintained in a population if another strategy is slightly better.

Using the game theory approach, it turns out that in many conflicts the evolutionarily stable strategy for fighting will not be one of outright aggression, but will involve a certain amount of formal display and ritualised combat. To see why this is so we will examine a simple game between individuals who adopt one of two imaginary fighting strategies – "Hawk" or "Dove". Hawks always fight hard and retreat only when seriously injured. Doves never fight, but merely display at each other threateningly and retreat if attacked. When a hawk meets another hawk both fight hard and each is equally likely to win the contest or be injured. When a hawk meets a dove, the hawk always wins the contest and the dove retreats unharmed. When two doves meet, nobody gets injured but both go on threatening each other until one gets tired and decides to back down.

We can now allot the contestants points which represent the benefits or costs for a particular outcome. Let us allot 40 points for a win, 0 for losing, -160 for a serious injury and -10 for wasting time in a long threat display. We imagine that these points can be converted directly into the currency of gene propagation into future generations. Box 1 shows the consequences of all four possible types of encounter in a "payoff matrix". Each cell in the matrix shows the average pay-off (number of points) for the encounter in question. For example, a hawk fighting another hawk will either win (obtaining 40 points) or suffer a serious injury (losing 160 points). Thus the *average* pay-off to a hawk fighting another hawk is -60 points.

How would evolution proceed in the hawk–dove game? First, consider what would happen if all individuals in the population played dove. The pay-off matrix dictates that all individuals would receive 10 points and nobody would be injured. Such a "conspiracy of doves", however, would not be stable, because if a mutation occurred causing an individual to play hawk, it would rapidly spread through the population because hawks playing doves obtain 40 points. In other words the strategy "always play as a dove" would not be an ESS.

Hawkish individuals, however, would not take over the entire population. In a population composed entirely of hawks, the average pay-off would be -60 points. A mutant dove invading this

1 THE HAWK–DOVE GAME

Points to: winner +40   injury −160
               loser   0   display −10

		Opponent	
		Hawk	Dove
Attacker	Hawk	$\frac{1}{2}(40) + \frac{1}{2}(-160)$ $= -60$	+40
	Dove	0	$\frac{1}{2}(40-10) + \frac{1}{2}(-10)$ $= +10$

Let the proportion of Hawks be h

The average pay-off to a Hawk, $\bar{H} = -60h + 40(1-h)$

The average pay-off to a Dove, $\bar{D} = 0h + 10(1-h)$

At equilibrium $\bar{H}=\bar{D}$; solving for h gives $h=\frac{1}{3}$, and therefore $(1-h)=\frac{2}{3}$. Thus the evolutionarily stable strategy is to play as a Hawk for one-third of the time and play as a Dove for the remaining two-thirds

population would do better than a hawk because when a dove meets a hawk it gets 0 (not a very high score, but still better than −60 for hawk–hawk fights). Therefore the strategy "always play as a hawk" is not an ESS either.

The essential feature of the hawk–dove game is that the best thing to do depends on what everybody else in the population is doing. Neither strategy can be stable when it is the only one in the population: the only evolutionarily stable strategy in this game is a "mixed" one in which individuals play sometimes as hawks and sometimes as doves. The stable mixture of the two strategies is reached when the points gained by playing hawk equal the points gained by playing dove so that there is no tendency for one strategy to displace the other. For the pay-off matrix shown in Box 1, for example, the population is at its stable state when individuals play as hawks for one-third of the time and play as doves for the remaining two-thirds. The implication, then, is that even though animals may be acting in their own selfish interests, natural selection does not always favour aggressive behaviour; a large element of doveish behaviour can be maintained in populations under some circumstances.

But to put too much emphasis on the restrained nature of animal fighting would be misleading. It is true that animals *do* sometimes inflict serious injuries and do sometimes fight to the death. The construction of evolutionary arguments in terms of game theory allows one to explore under precisely what circumstances serious fighting will be favoured. By manipulating the costs and benefits of the different options – the points in our pay-off matrix – it is possible to show that some populations could evolve in which animals *always* fight like hawks. This, however, is likely to occur only when the value of a prize far outweighs the cost of fighting for it, so that it will always be better to fight than display. This means that the strategy "always play as a hawk" could be stable sometimes, especially when animals are fighting for very valuable resources.

To males of polygynous species, harems containing a large number of females constitute highly prized resources. This is because a male who possesses a harem can potentially fertilise all the females as they come into oestrous, and so ensure his paternity of the offspring. Not surprisingly, then, we find that in harem-holding species such as elephant seals and red deer, males fight in earnest, often inflicting serious injuries which can sometimes be fatal. Fights between male elephant seals are dramatic contests of strength in which opponents rear their huge bodies up off the ground and lunge at each other inflicting deep scars with their sharp teeth. Stag fights in red deer are prolonged pushing contests in which the contestants interlock their antlers and twist them from side to side; injuries include broken antlers, broken legs, and permanent blindness. In Russian red deer, rutting injuries account for about 20 per cent of all adult male deaths, while in German populations, 5 per cent of stags die every year by fighting.

Although the hawk–dove game can be used to make predictions about the circumstances under which aggression will and will not be favoured, it is obviously a very simple representation of how we can expect real animals to behave. It is simple because it assumes that all the "players" are identical. In many species, individuals often differ in age, body size or experience, and all these things are likely to affect fighting ability. In real populations, smaller, younger or less experienced competitors usually avoid serious fights and make the best of a bad job by adopting sneaky strategies. In red deer, for example, smaller younger males (called sneaky rutters!) do not attempt to fight with stags to gain possession of harems but settle for

*Many species compete for territory or mates, including zebras,
elephant seals, red deer and great tits*

occasional copulations with females when the harem-holding stag is not looking. As a countermeasure against sneaking, stags attempt to drive these young males away from the harem but sometimes they are unable to do so, perhaps because they are defending the harem from another stag.

How should fights be settled between individuals who differ in fighting ability? This question can be explored by inventing new strategies and introducing them into the hawk–dove game. One strategy we can introduce is called "assessor". Assessors, as the name suggests, spend some time assessing the size (or other indicator of fighting ability) of their opponent before any combat begins. The rule they play to is "if larger than opponent, play as a hawk; if smaller, play as a dove". Intuitively, this strategy sounds sensible, and it often turns out to be the ESS in a game played between hawks, doves and assessors.

In nature, assessment commonly occurs prior to fighting and opponents often respect differences in fighting ability as a means of avoiding escalated battle. Red deer stags engage in prolonged roaring contests to gauge each others strength and stamina before getting down to serious combat. Many stags back down after a roaring contest, presumably because they judge themselves to be weaker than their opponent and less likely to win in a fight (see p. 50). Similarly, male common toads (*Bufo bufo*) assess the body size of their opponents before fighting over females. They do this by listening to the pitch of their opponents' croaks.

Although I have so far emphasised the application of game theory to the evolution of fighting behaviour, the theory has been fruitfully applied in many other contexts (reviewed in Maynard Smith's book *Evolution and the Theory of Games*, Cambridge University Press, 1982). It has been shown to be useful in thinking about how parents should allocate their resources between male and female offspring and, in the context of animal communication, whether individuals should signal information about their intentions.

Game theory also has widespread applications to sexual rivalry. Intense competition between males for mating opportunities is evident in many species, though this does not always take the form of overt aggression. Modern textbooks on animal mating strategies, although intended for serious reading, often read like the *Kama Sutra*. There are male mites that mate with their sisters before they are born, male insects that use chastity belts to block up the genitals of their mates, male bugs that commit homosexual "rape" and inject sperm into the bodies of other males, and many more.

G

In some species, males even resort to mimicking females in an attempt to obtain clandestine copulations. The mating strategies of male scorpion flies (*Hylobittacus apicalis*), studied by Randy Thornhill at the University of New Mexico, provide a fascinating example of this. Usually, male scorpion flies capture small insects and offer them to females as prenuptial gifts prior to copulation. But catching insects is a risky business because it is easy to become trapped in a spider's web while hunting for prey. To avoid becoming a snack for spiders, some males steal the prey caught by other males. This is done by flying at a male and knocking the prey from his grasp, or by using the more devious method of posing as a female to try and dupe the other male into giving up his prize. Although this "transvestite" strategy is often successful, the victim of the robbery sometimes grabs his prey back when he realises the other individual refuses to copulate!

Just as in the hawk–dove game, it is easy to see why the two strategies hunting and stealing may be maintained in the scorpion fly population. As more individuals hunt for prey, it becomes more profitable to steal. It is obvious, however, that all males cannot steal all the time because there would soon be no more prey to steal. The pay-offs to the two strategies hunting and stealing are negatively frequency-dependent – one strategy will always be favoured when the other is common. Game theory would predict that the population of scorpion flies should evolve to consist of a stable mixture of hunters and stealers with both strategies doing equally as well.

Thornhill worked out the success of the hunting and stealing strategies for scorpion flies and came to the surprising conclusion that stealing was a more profitable way of obtaining mates than hunting. How, then, can this result be reconciled with the game theory prediction? One possibility is that scorpion flies are playing a conditional strategy (similar to assessor) which specifies, If large, steal; if small, hunt. Maybe just as only the largest or strongest red deer defend harems, so only the largest or strongest scorpion flies go in for stealing. Less competitive flies may be forced to make the best of a bad job by hunting.

The application of game theory to animal behaviour has not been restricted only to competition; it is also helpful in understanding cooperation and altruism. Cooperation among animals presents an even greater problem for Darwinists than does ritualised aggression. Darwin himself realised that his theory of natural selection would be disproved if it could be shown that animals behaved altruistically solely to ensure the survival of another individual. Yet associations for mutual benefit appear to be common in the animal kingdom. Cooperation often occurs between closely-related individuals, such as parents and offspring. This kind of cooperation presents no problem for the theory of evolution because individuals are helping to ensure the survival of other individuals whose bodies contain copies of their own genes. But, as the degree of relatedness between individuals decreases, it becomes less likely that two individuals will share copies of the same gene for cooperation and individuals would do better to look after their own interests instead. Therefore, the plentiful observations of unrelated animals (even different species!) cooperating to defend territories or nest sites, or to look after young, presents something of a problem for evolutionary theory.

Robert Axelrod and William Hamilton explored the possibilities for an evolutionarily stable strategy of cooperation between non-relatives using the familiar prisoner's dilemma game (see *New Scientist*, vol. 97, p. 432). In this game there are only two strategies: cooperate and defect (see Box 2). From the pay-off matrix it can be seen that if player A cooperates, it pays player B to defect; and if player A defects, it pays player B to defect also. Therefore, if individuals play only one game against each opponent, the only ESS is "always defect".

---

2 THE PRISONER'S DILEMMA GAME

Each player has the choice to Cooperate with the other or Defect
The matrix shows the pay-offs to player B only

		Player A's choice	
		Cooperate	Defect
Player B's choice	Cooperate	30 = reward for mutual cooperation	0 = sucker's pay-off
	Defect	50 = reward for defection	10 = punishment for mutual defection

Conclusion: no matter what player A does, it always pays player B to Defect

Things are quite different, however, when individuals play several games against each opponent; another strategy called "tit-for-tat" can also be an ESS. Individuals who play tit-for-tat always cooperate in the first game and in all subsequent games they make a choice dependent on what their opponent did in the previous game. If in the last game the opponent cooperated, then tit-for-tat responds by cooperation; if the opponent defected, then tit-for-tat defects. The tit-for-tat strategy is highly successful when played against other strategies in computer simulations; in fact it accumulates more points than the strategy "always defect". When the strategy tit-for-tat is present in the population, no other strategy does better: tit-for-tat is therefore an ESS.

Is there any evidence that real animals use the tit-for-tat strategy when cooperating with others? Craig Packer studied the behaviour of olive baboons and found that males often form partnerships to fight for females against other males. Normally, receptive females are guarded by a single dominant male who attempts to maintain exclusive mating rights. Subordinate males have little chance of taking on the dominant male singlehandedly, so they form fighting coalitions by enlisting another male to help fight against the dominant. While the enlisted male distracts or fights the dominant male, the other male copulates with the female. Then, when the next occasion to fight for a female arises, the two subordinate males reverse roles and the male who previously copulated now fights the dominant male while the other one copulates. This system of reciprocal altruism seems to be a good thing for both males: by taking turns at fighting and copulating both males obtain more matings than they could possibly achieve alone.

The crucial question is what prevents males from cheating? It seems that a male would get the best of both worlds if he enlisted other males to help fight for females but failed to return the favour on future occasions. The reason that males do not cheat is probably that baboons can recognise each other individually and can bear a grudge against cheaters by refusing to cooperate with them in the future. In other words, males who always defect would be punished on future occasions by males who play tit-for-tat.

Although the theory of games and the concept of the evolutionarily stable strategy has greatly helped biologists to understand animal behaviour, it does have its limitations when applied to the real world. We can go on imagining different strategies *ad infinitum* but it would be unrealistic to assume that animals can follow any strategy that we care to include in our models. For example, an animal with a machine gun might do very well in the hawk–dove game, but until we see real animals wielding machine guns there is little point in putting this strategy into the game. Clearly, we must know which strategies are feasible before constructing detailed models of behaviour.

Population geneticists have criticised the use of game theory on the grounds that it does not tell us how populations change with time; it merely describes the population at its equilibrium state. The game theory approach, however, provides a framework within which a wide range of phenomena, from fighting and aggression to cooperation, can be discussed. It also enables us to make fairly precise quantitative predictions about behaviour – such as when to fight, how long to persist in a struggle, and when to cooperate – without a detailed knowledge of the genetics of populations.

25

Animals make the most of the home-field advantage

KEN YASUKAWA

1 November 1979

The criteria animals adopt to decide the outcome of a squabble may be quite arbitrary, such as whether or not the animal is on its home ground.

Few sports enthusiasts – participants and fans – doubt the importance of the "home-field advantage". The support of a vociferous crowd and the psychological benefits of playing at home can improve individual and team performances. Tournament games in several sports are often played on a neutral field to minimise this advantage. In a sport such as golf, where courses vary considerably, the advantage probably results, in part, from a knowledge of the peculiarities of the home field. However, in a sport such as soccer, where playing fields vary little, the advantage seems more psychological. Biologists have recently started to think of encounters between animals as analogous to sporting contests in many ways. For example, early studies of animal behaviour seemed to demonstrate home-field advantage.

Lesser black-backed gulls (*Larus fuscus*) defend adjacent, apparently identical, nesting territories, and often engage in aggressive encounters. When bird A intrudes on bird B's territory, bird A invariably loses the encounter. However, when bird B is the intruder, A defeats B. Amazingly, this "resident-wins" phenomenon can override differences in actual fighting ability. Even when the intruder is an adult and the resident is only a juvenile, the young resident wins. Under other circumstances the older bird could easily defeat the youngster but residence seems to produce effects similar to the home-field advantage in sports.

Home-field advantage in animal tournaments has received attention from theoretical biologists such as John Maynard Smith and

Geoffrey Parker. They see an encounter between two individuals as a contest that can be analysed using mathematical models based on game theory and our current understanding of evolutionary biology. In the beginning, these models assumed that contestants were essentially equal in all important respects, even though actual contests between real animals are very unlikely to meet this assumption. Subsequent models attempted to account for "asymmetries" that might exist between contestants. These asymmetries could result from differences in fighting ability, or differences in the potential benefit of winning, or other differences in characteristics that to us seem completely arbitrary. According to these newer models, if contestants could assess the asymmetries, they could use them to settle contests without recourse to combat. The animals would then use elaborate rituals to assess their relative abilities and to resolve their differences without actually fighting.

Maynard Smith and Parker argued that if one contestant were a better fighter, or stood to gain more from a victory, or to lose more from a defeat, he or she would probably be more willing to fight. This willingness to escalate an encounter may explain why differences in fighting ability and potential benefit affect the outcome of a ritualised contest. However, it is difficult to understand why some arbitrary asymmetry, analogous to the toss of a coin, should be used to settle disputes. A real asymmetry would be, for example, a difference in muscular strength. An arbitrary asymmetry can be any difference between the contestants that does not involve fighting ability.

Male speckled wood butterflies (*Pararge aegeria*) use an apparently arbitrary rule – the resident always wins – to settle disputes over patches of sunlight on the forest floor. Nicholas Davies, at Oxford, did the experiments that show that the occupant of a sunlight patch invariably wins an encounter, and that the outcome of a contest between two individuals could be reversed by only a few seconds of residence in the patch by the former loser, a clear home-field advantage.

I was studying ways that previously unmated male redwinged blackbirds (*Agelaius phoeniceus*) go about establishing their territories when I observed a phenomenon that might be explained by the effects of asymmetries. Males who successfully acquired a breeding territory appeared to be less aggressive than unsuccessful males. As territoriality involves aggressive defence of an exclusive area, I was surprised to find that the successful males were *less* aggressive than unsuccessful ones. This result suggested to me that successful males possessed some other advantage that allowed them to establish their territories more easily. Searching through my records, I found that males who had greater opportunities to gain experience a year or more before they actually established territories were the most successful. This experience may have produced some advantage, or asymmetry, that increased the probability of the male getting a territory and reduced the aggression necessary for him to do so successfully. Unfortunately, I did not have enough data to identify the exact nature of the asymmetry or how it affected aggression.

My research on how redwinged blackbirds establish territories led me to seek out a species with which I could experimentally manipulate experience. I chose the dark-eyed junco (*Junco hyemalis*) because earlier studies had shown clearly that body size, age, plumage darkness, sex and length of tenure in an aviary could affect the rank of a bird in winter flocks. Larger, older, darker males who had been there longer were likely to be the top birds. Since the effects of these non-arbitrary asymmetries, which reflect fighting ability and potential benefit, seemed well understood, I focused on arbitrary asymmetries by conducting experiments designed to answer two questions: first, can arbitrary asymmetries determine dominance status? Secondly, can arbitrary asymmetries reduce the aggression necessary to establish dominance?

Any study of arbitrary asymmetries must first take care to minimise the effects of non-arbitrary differences, so I ensured that all experimental subjects were large (wing length >81 mm) males, and all were dark-slate coloured. In experiment 1 the birds were all older than one year, while in experiment 2 they were all younger

*The dark-eyed junco (*Junco hyemalis*)*

than one year. As well as matching the birds for physical character-istics, I tried in addition to ensure that they had the same experience in the aviaries, so all the birds were given a week to become familiar with one of four large aviaries that were as identical as I could make them. I also assumed that the dominance rank a bird reached during the week of familiarisation might affect his willingness to fight, so I made comparisons only within high ranking birds or within low ranking birds.

All these tedious procedures were very necessary to make sure that I had removed as many differences as possible. If I had been successful then an arbitrary asymmetry might affect not only the amount of aggression shown by the birds in establishing their hierarchy, but also the position of individual birds within the hierarchy. The arbitrary asymmetry I had in mind was, of course, the home-field advantage.

In experiment 1, six large, dark, old males were placed in aviary A, and six others in aviary B. All were given one week to establish their hierarchies and become familiar with life in the aviary. After the week of familiarisation, all 12 birds were removed, and a subgroup of the three high ranking birds from aviary A was simultaneously reintroduced back into that aviary along with the three high ranking birds from aviary B. These six birds constitute the high rank, home-away group (H-H). The remaining six birds were simultaneously introduced into a third aviary, C. These birds consti-tute the low rank, neutral-ground group (L-N).

In experiment 2, six large, dark, young males were placed in aviary A, six others in aviary B and given one week to become familiar as before. A week later, in contrast to the first experiment, the two high ranking subgroups were placed in a third aviary, while the remaining low ranking birds were placed in aviary B. These constitute the high rank, neutral-ground group (H-N) and the low rank, home-away group (L-H), respectively.

To see if prior residence is indeed an arbitrary asymmetry that can determine status and reduce aggression when male dark-eyed juncos form a dominance hierarchy, I tested the following predictions:

(1) Birds "at home" should become dominant over birds "away", but because individuals within subgroups had established dominance relationships during the familiarisation period, their prior ordering should be retained.

(2) Birds in neutral aviaries should retain any prior dominance

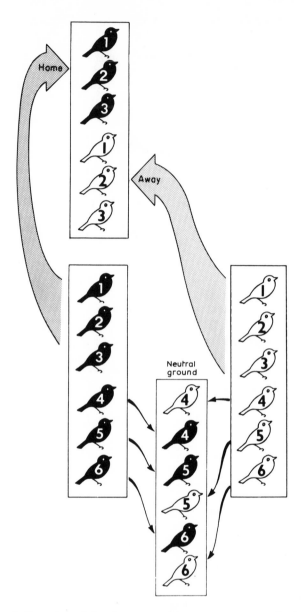

Figure 1 *Results of the first experiment were as predicted. Each group's hierarchy was stable, so that the order of the birds stayed the same, but the high-ranking birds who stayed in their home aviary were able to dominate the other high-ranking birds who were in an unfamiliar aviary. Where neither group had the home-field advantage, as in the low-ranking birds, the two hierarchies mingled*

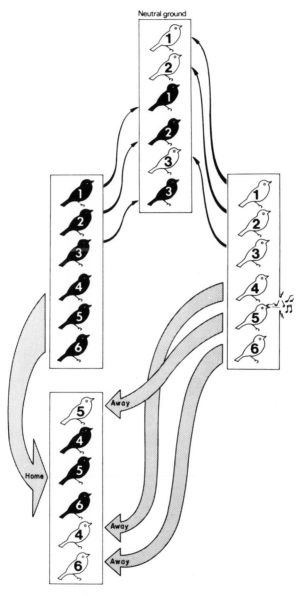

Figure 2 *Results were also as predicted with low-ranking birds for the home-away match, except for away bird number 5. He started singing early, and went to the top of the ladder*

relationships, but because neither subgroup has the home-field advantage, their hierarchies should mingle.

(3) Regardless of rank, the home-away groups should be less aggressive than the neutral-ground groups because the former can use an asymmetry to settle encounters.

With the exception of one of the younger birds, to whom I will return later, all my predictions came true. Figures 1 and 2 show the results in graphic form. All subgroups retained their prior order, so that the hierarchies that had been established in the week before the contest were stable. In the neutral aviaries, the two subgroups did indeed mingle as they formed the new hierarchy, whereas in the home-away aviary, the home team was always dominant to the visitors.

Furthermore, Figure 3 shows that home-away birds, who could use an arbitrary asymmetry to settle their differences, were less aggressive than birds in a neutral aviary. They threatened more, but the threats did not escalate to actual fights nearly as often as in the neutral aviary. This difference was more marked among the high ranking birds, but was still present in the low ranking birds.

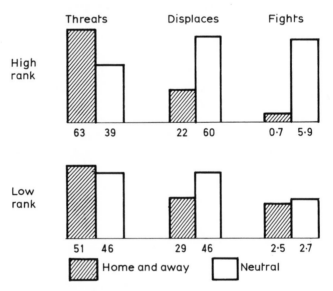

Figure 3 *The number of threats, displacements and fights in a half-hour session. Birds with an asymmetry – the home-away distinction – threatened more often, but these threats did not escalate into real fights*

These data indicate that an experimentally controlled, arbitrary asymmetry, in this case prior residence, can affect the aggressiveness of dark-eyed juncos, and the outcome of contests between them.

Bird B5 in experiment 2 was an interesting exception because his behaviour might be explained by an asymmetry in fighting ability that I had not controlled. Young male dark-eyed juncos, like many song birds, undergo a series of changes that prepare them for their initial return to their breeding grounds. To accomplish the tasks of acquiring a territory and a mate, the young male must produce songs typical of the species and aggressively exclude other males from his territory. Song development progresses from so-called subsong through plastic song to full song, and is accompanied by increases in testes size, testosterone levels and aggressiveness (*New Scientist*, vol. 82, p. 536). Male B5, the apparent exception to my predictions, was the only young male who produced subsong during the one-week familiarisation period. Although other males began to produce subsong during the experimental period, B5 sang more frequently than any other member of the low rank home-away group. This song development was accompanied by increasing aggressiveness as B5 began to exhibit territorial behaviour. Although these changes occurred too late to affect the status that B5 attained during the familiarisation period, they may have created an asymmetry in fighting ability during the experimental period. My experimental procedure could not control an asymmetry resulting from the development of song and territorial behaviour, and these may have allowed B5 to become dominant. If these speculations are correct, they indicate that non-arbitrary asymmetries can override the effects of arbitrary asymmetries. This possibility will require further study but it seems clear from my experiments that the home-field advantage can be an important factor in resolving animal contests.

26

Animals as gamblers

TIM BEARDSLEY

2 June 1983

Few animals know for certain where their next meal is coming from. If they look in the wrong place, there is a real chance that they will starve. But in practice animals seem to be rather good at coping with life's uncertainties.

Vladimir: That passed the time.
Estragon: It would have passed in any case.
(S. Beckett, Waiting for Godot, Act 1).

Estragon's bizarre remark touches on a fundamental problem for animals. To be more accurate, the problem has been for those trying to understand behaviour as a product of natural selection, since the animals themselves seem to have been coping quite well with life's ups and downs for a long time. The problem is that of relating *fitness*, which depends on behaviour over a whole lifetime, to behaviour that is observed in the short term. Any evolutionary explanation of behaviour has ultimately to be in terms of its contribution to fitness (see Box 1). But fitness also depends on chance: in an animal's natural environment there is no certainty about the outcome of any action. An animal never knows for sure what is coming round the next corner. How then are we to assess each of the moment-to-moment decisions that animals make throughout their lifetimes? Or to return to Beckett, why should Vladimir and Estragon wait for someone who might not turn up, rather than look for more carrots?

Evolutionary (or *functional*) explanations of behaviour are nothing new: Niko Tinbergen was testing them in the 1950s. With this approach ethologists do not attempt to say anything about *how* the mechanisms in an animal's head make it behave the way it does,

---

1 BEHAVIOUR AND EXPECTED FITNESS

As there is genetic (heritable) variation between individuals that affects behaviour, natural selection should favour genes that are effective at predisposing their carriers to increase the genes' representation in subsequent generations. An animal may contribute its genes to succeeding generations by rearing offspring of its own and by helping others of the species that are related to it to reproduce. The genes' frequency in some later generation is a measure of the animal's *fitness* (more properly, its *inclusive fitness*, but as we are not talking here about interactions between individuals we can drop the "inclusive" and just talk about fitness). Behaviour that tends to maximise *expected fitness* (the average of each possible fitness outcome multiplied by its probability of occurring) is called *adaptive*. Although behaviour ought to be adaptive, there are a number of reasons why a particular sequence of behaviour might not be perfectly adaptive: the most important is that if the environment is very different from that in which the animal evolved, its decision-making mechanisms might be "fooled".

---

but attempt to explain *why* an animal benefits in evolutionary terms from behaving in the way it does.

To do this properly the investigator has to compare the consequences of different courses of action. Is it better, for example, to walk 50 m to drink, than to walk 100 m to eat, or to court a possible mate? The problem is complicated when we realise that the animal is getting hungrier and thirstier all the time and that, if it chooses water, the water hole might be dry and both food and mate gone by the time it got back. Recent theoretical work offers the hope of accounting for behaviour when faced with problems such as this in terms of its contribution to fitness. The evidence so far – and there is precious little of it – is that animals are rather good at gambles of this sort.

Mathematical models based on optimality theory have been widely used in recent years to predict what decisions would be made under different conditions by an animal that was continuously maximising its expected fitness. Such models need two things, in addition to the requirement that expected (average) fitness should be maximised at all times. The first is a list of the constraints that limit behaviour, for example, the maximum rate at which the animal can feed. These constraints are not usually known, but sensible estimates can often be made – in this case, perhaps "the fastest feeding rate we are able to observe" would be used. The second thing optimality theory requires is that we specify – in some form – the relation between any decision that could be made, and the resulting expected fitness. This is where the trouble really starts.

The fitness/decision relations are simply not known. Ethologists have had to proceed by making educated guesses about what the relations might be. Nevertheless, they have had considerable success, notably in predicting what sort of food a foraging animal should take, when it should feed and drink and when it should move between different local concentrations of food. The successes provide support for the constraints and the decision/fitness relations incorporated into the optimality models, and also provide support for the notion that foraging decisions are to some extent adaptive. As animals are not omniscient, they can be expected to behave adaptively only if the decision mechanisms in their heads will "work" in the context being investigated: for this reason the circumstances under which the animal is tested should be as similar as possible to the sort of environment in which it evolved.

But optimality models that have been applied to foraging decisions also consistently fail to predict some aspects of observed behaviour. This does not mean that the behavioural decisions involved are maladaptive: such a verdict would be possible only if all the constraints and the fitness/decision relations were known precisely. It does mean, however, that the simplifying assumptions in the models have to be scrutinised.

Most of what follows is about the sort of decisions an animal ought to make to maximise its expected fitness, given the constraints on its behaviour. I will use the word "optimal" for choices that maximise expected fitness. Real animals cannot calculate optimal choices any more than the dogs in television advertisements can analyse the chemical difference between "Doggo" and Brand X, but the dogs can still apparently tell the difference and make a choice. Real animals might often make optimal decisions from the same position of blissful ignorance. If their decisions consistently maximise expected fitness, this is evidence of adaptation, however the decisions are made; there is no need to infer foresight or understanding on the part of the animal. Animals in fact often appear to use simple rules which approximate optimal behaviour under appropriate conditions.

The evolutionary reason why an animal survives is in order to reproduce (or to help related individuals to survive or reproduce). It is usually supposed that an animal behaving optimally would maximise expected reproductive success, though this is not always true (see Box 2). Optimality analyses of how much effort an animal should spend on reproduction and when it should reproduce have also been developed, but I will not discuss them here. This is because the impact of uncertainty on optimal decision-making is most easily explained by considering just the effect of feeding decisions on probability of survival. Similar effects will apply to many sorts of decisions, however.

For the sake of simplicity, imagine that there is only one kind of food (which occurs as discrete items) and that the animal has no predators. One commonly-assumed function relating foraging decisions to fitness is that expected fitness is maximised by maximising

---

2 EXPECTED FITNESS AND EXPECTED REPRODUCTIVE SUCCESS

It is often assumed that maximising expected reproductive success (usually taken as the number of offspring that survive to adulthood) is the same as maximising expected fitness. But John Gillespie at the University of Pennsylvania has shown that to maximise expected fitness it is necessary to take account of the variance (uncertainty) in the number of offspring. For example, animal A and animal B have identical expected reproductive successes:

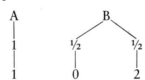

Animal   A   B

Probability   1   ½   ½

Offspring   1   0   2

The expected frequency of A-type animals in the second generation (a measure of A's fitness) is

$$\left(\frac{1}{2} \times \frac{1}{1}\right) + \left(\frac{1}{2} \times \frac{1}{3}\right) = \frac{2}{3}$$

whereas the expected frequency of B-type animals is only

$$\left(\frac{1}{2} \times \frac{0}{1}\right) + \left(\frac{1}{2} \times \frac{2}{3}\right) = \frac{1}{3}$$

When population size is approximately constant (which it is not here) the effect is small, unless the uncertainty for B-type animals is between different generations, as distinct from within a single generation as here. Gillespie's principle is quite distinct from the other uncertainty problems in this article, but it leads to the same sort of prediction: animals should distrust the uncertain outcome.

the average rate of food intake (MARFI) over the period the animal is foraging. This convenient measure for fitness has limitations, which are now becoming apparent. The limitations are classed into three groups: limiting future options, hunger and survival, and learning.

Limiting future options

It is possible for a strategy that is MARFI in the short term to be not-MARFI in the long term simply because it affects foraging options that might have occurred later. If you were stranded for two days with no food except for a hen, you would probably eat the hen. But if you anticipated being stranded for two weeks, it would be better to keep the hen and live off its eggs, even though you would eat less during the first two days than if you ate the hen. It is hard to know how important this problem is but it may be significant for animals that defend territories, or for parasites. The general message is that MARFI is not expected from an animal behaving optimally if future foraging options have been altered over the foraging period in question.

A similar effect could explain the fact that pigeons in Skinner box experiments usually show a strong "time preference" for rewards that come immediately over rewards that are delayed, even by a few seconds. This may result in their getting a lower average rate of food intake than they could achieve by waiting for a better reward. But if pigeons are adapted to feeding in the presence of competing individuals, food that is not snapped up as soon as it is spotted might have gone 30 seconds later. A pigeon seems to gamble on the assumption that if it does not eat the food now, someone else will – a safer bet than waiting for something better to come along.

Hunger and survival

A MARFI-behaving animal can go badly wrong if the state of its body food reserves affects its survival prospects. In this case, even if all items of food are the same size, they do not all have the same value. To make another human analogy, you would pay a lot more for a sandwich if you were about to faint from hunger than you would just after finishing a three-course meal. This is why uncertainty is important for an optimal decision-maker. Consider a bird choosing between two foraging options: choice A is certain to yield 2 units of energy from food, whereas choice B is just as likely to yield 4 units or nothing at all – each outcome has a probability of ½. A MARFI animal would be indifferent to both, since the expected (average) energy gain from B is ([½ × 4] + [½ × 0] = 2) is the same as that from A (1 × 2 = 2). But what if the bird uses up 2 units of energy during the choice test, and has only 1 reserve energy unit stored in its body at the start of the test? In this case, choice A guarantees the bird's survival, whereas with choice B it has a chance of ½ of starving to death, when its reserves go below 0.

This is a simplistic example but it illustrates the general point that, because an animal's state of hunger matters for survival, the uncertainty associated with a choice matters, as well as its expected yield. In this example an animal behaving optimally would favour the certain outcome A over the uncertain outcome B. Preference for the less variable outcome when two expected yields are equal is known as *risk-averse* behaviour.

In this admittedly extreme example, the optimal choice is rather obvious. If the bird were still several energy units away from starvation before the test, and so in no immediate danger, predicting the optimal choice would be more difficult, because we would not know how to evaluate the alternative energy states the bird might end up with. John McNamara and Alasdair Houston have shown a way out of this problem: in essence, their idea is that the contributions to fitness of the two options could be evaluated by looking at their effects on survival over the period from the end of the choice test to some point in the future. If the food intake has a random component – as it would if the bird were faced with a series of tests like this one – that point need not be very far in the future.

By using this approach, McNamara and Houston have argued that the function relating expected fitness to the level of an animal's body food reserves at the end of a foraging session will usually take the form of a sigmoid curve, as shown in Figure 1. The exact shape of the curve does not really matter; the point is that for many different detailed models the curve will be this sort of shape if chance has a part to play when an animal finds food – which is, of course, just about always.

It follows from a mathematical result known as Jensen's Inequality that where the curve in Figure 1 is convex (region A) an animal should favour *uncertain* outcomes when other things are equal. This is *risk-prone* behaviour. The reason can be illustrated with the bird example used above: if the bird lost 4 units of energy during a choice test, rather than only 2, its only hope would be to take choice B instead of A. Choice A would mean certain starvation. Where the curve in Figure 1 is concave, on the other hand (region B), an animal behaving optimally would be risk-averse.

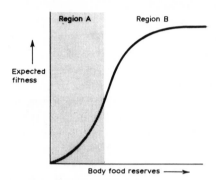

Figure 1 *When body food reserves are low, expected fitness is also low, and the animal is forced to take risks (region A). An animal lucky enough to be in region B of the curve is wiser not to take chances*

The message from all this is that, because degree of hunger affects survival, we expect animals to display risk-averse behaviour when they are comparatively well stocked with energy reserves, and risk-prone behaviour only when they are in dire straits. Animals should be rather conservative gamblers most of the time and prefer the safe option to the risky bet, because it is more important not to do very badly than it is to do very well.

So much for the theory. How do real animals shape up? Tom Caraco and others performed an experiment at the University of Arizona which examined this question. They presented yellow-eyed juncos (North American birds of the bunting family) with two choices. The first alternative was to take n seeds with probability 1; the second was a gamble consisting of $n-k$ seeds with probability $\frac{1}{2}$ or $n+k$ seeds with probability $\frac{1}{2}$. The expected (average) number of seeds from the two choices was therefore equal. The results were just what would be expected from the ideas above: the birds showed risk-averse choices when they were not very hungry and allowed to feed quickly, but risk-prone choices when more hungry and constrained to feed slowly. This supports the idea that the state of hunger affects fitness and suggests that yellow-eyed juncos are adapted to cope with uncertain outcomes. Or we could put it more romantically and say they are good at gambling with fate.

Myron Baker and his colleagues at Colorado State University showed that dark-eyed juncos also know what is best for them. Socially subdominant individuals seem to gain little from feeding in flocks, as they are constantly being robbed of food by more dominant birds. But though their average rate of food intake does not go up, the probability that they will starve goes down. Flocking decreases the chance of getting nothing at all to eat, and so this strategy turns out to be optimal even though competition with others might reduce the maximum that a bird could obtain.

Learning

There is another way in which animals are adapted to cope with uncertainty, which is quite different from the idea above. Hungry animals choosing between different foraging areas ("patches") typically spend most of their time in the patch that allows them the highest rate of food intake. But they also persistently spend at least some of their time foraging in inferior patches. This is not because they cannot tell the difference, or forget. Because they do this they get a lower rate of food intake over a foraging session than they could if they spent all the time in the best patch, so it is not MARFI behaviour.

Numerous experimenters have found this sort of effect in their data, using many different species. The explanation usually offered is that the subjects are *sampling* different patches so that should relative food intake rates in the patches change suddenly, it would be immediately clear where the new best patch was.

If this explanation is correct, it could represent an adaptation to cope with varying qualities of patches by *learning* about how good the different patches are. The subjects are gambling on the possibility of some change in the quality of the patch. As long as the patches do not change, the gamble does not pay. But if the experimenter does rearrange the patches, the subjects quickly identify the new best patch. It is not hard to imagine that this tendency might be adaptive in the wild, where rates of food supply in different places are continually changing.

Of course, it will not do just to invent a good explanation like this. They could be moving between the patches simply for the exercise, unlikely though this is. The sampling idea has to be tested. This means, as before, that the advantage of sampling has to be demonstrated in terms of its contribution to fitness. When it learns, the animal acquires information that could be useful later. The contribution to fitness of each attempt at sampling should, therefore, have a term for the value of information acquired as well as any food that may result. Experiments have been performed to see if animals sample different patches only when this is optimal (see Box 3). Results suggest that animals sample in such a way that expected fitness *tends* to be maximised, even if this is not achieved perfectly. It is possible to think of the experimental subjects as experimenters themselves, testing hypotheses about feeding rates.

---

3 SAMPLING AND EXPLOITING

Consider the problem of a hungry bird faced with two levers that occasionally release food rewards. Each lever gives a food reward with a fixed probability when pressed, but the two probabilities are different. Clearly an animal behaving optimally would feed exclusively on the better lever if it knew which this was. How is it to find out? It has to spend some time *sampling* both levers, before settling down to *exploit* the best one. Real animals do this too.

How much time should be spent sampling as opposed to exploiting? The problem is that the longer a bird spends sampling, the longer it is feeding more slowly than it could on the better lever alone. On the other hand, the longer it spends sampling, the more sure it can be about which lever is really the best.

It turns out that the optimally behaving animal should spend longer sampling if the foraging session lasts a long time than if it lasts a short time. Alex Kacelnik gave this problem (called a two-armed bandit problem) to great tits, and found that they do indeed spend longer sampling during a long foraging session than during a short one. This suggests that sampling is an adaptive strategy which animals tend to use to monitor food availability only when to do so is optimal.

The evidence, then, suggests that animals are good gamblers, even though they may not be perfect in any given experiment. It also shows that employing optimality analysis can lead to new insights into why animals behave the way they do. We should soon be able to give Vladimir and Estragon some good advice.

PART FIVE

Understanding the Human Animal

Darwin's contemporaries were scandalised at his assertion that man and other animals shared equally lowly ancestors. The full implications for human society of man's non-human origins are still far from clear. Is most human behaviour biologically determined, or has the evolution of culture enabled us to break free from the tyranny of our genes? If the latter is true, there is little sense in looking to animals to enlighten us about our motives (or rather, the motives of our genes). If, as seems more probable, human behaviour does follow at least some of the same rules as the behaviour of animals, we may be able to learn something from them.

At times man has been too eager to turn to animals for easy solutions to the problems that beset his society. During the 1950s and 1960s, almost every article on animal behaviour had a prim concluding paragraph in which morals were drawn for mankind. The late David Lack countered this tendency by pointing out (p. 201) that although birds rarely fight to the death, only the strongest of them will have access to the resources necessary to survive – hardly a model for a caring society.

Man is unlike any other animal in his predilection for killing in large numbers, whether his own or other species, for reasons unconnected with food. Some carnivores also, on occasion, go in for mass slaughter (p. 206); if their prey fails to take evasive action, they seem not to know when to stop. Man's superior technology for killing may put him in a similar position but the superior reason he reportedly possesses should be able to counteract such a failure of his instinctive braking mechanism.

The practice of drawing analogies between human and animal societies took on another dimension with the advent of sociobiology. This "new synthesis" is on the face of it no more than an attempt to produce functional explanations of social behaviour that apply as well to humans as they do to animals. The subject was effectively launched in 1975 with the publication of E. O. Wilson's book *Sociobiology*, in which he listed universals of behaviour that he saw throughout the animal kingdom, including man. He reasoned that such universals (avoidance of incest was one example) must be genetically determined (p. 212).

Although Wilson went to some lengths to warn against the "naturalistic fallacy" of assuming that because something exists it must be meant to exist, his views attracted intense hostility (p. 219). Opponents included radical scientists such as the Harvard biologists Steven Jay Gould and Richard Lewontin and sociologists. They regarded Wilson as a politically dangerous supporter of (at best) capitalist society and (at worst) of Nazi-style eugenics. So incensed were they that they adopted a position of extreme opposition and often resorted to distorting Wilson's thesis by quoting *very* selectively (and out of context) from his book.

Today some of these opponents recognise that they probably damaged their case by overreacting in the first instance. Most biologists now accept the broad outlines of Wilson's analysis, although they may disagree with him on the details. As Roger Lewin discusses on p. 226, sociobiology offers a framework within which to consider such knotty philosophical questions as, How does man come to have an ethical sense?

27

Of birds and men

DAVID LACK

16 January 1969

Birds often solve their disputes without actually coming to blows and almost never fight to the death. But death may well be the eventual outcome for the loser of he is denied access to food. If there is a lesson here for human society, is it really worth learning?

We are desperately anxious to stop wars, so it is scarcely surprising that ornithologists are often asked whether a fuller knowledge of aggression in birds would reduce warfare in mankind, the implied answer being yes. From reading such books as Konrad Lorenz's *On Aggression*, many people have learned for the first time how much of the fighting of birds and other animals is ritualised, and how many disputes are settled without bloodshed, or even blows. It seems an idyllic world, and hence it seems at first sight reasonable to ask whether the animal basis for human aggression might not be modified, or changed back, to a condition like that of birds and bring us peace. But this ignores the ecological results of fighting in both birds and man.

In birds, members of the same species fight each other for two main objects, food and breeding territories. Fighting for food is seen chiefly in winter among birds that feed in flocks on the ground. Flocking probably reduces the chance of a bird of prey approaching undetected, but it often enables one bird to take an item just found by another. In such cases the aggressor displays, usually mildly, the other bird retreats and the aggressor eats the food. In each encounter between two particular domestic fowls, the same individual retreats, and birds in a flock can be arranged in a peck order, from the one at the top, which displaces all the others, to the one at the bottom, which gives way to all the rest.

This happens because the birds recognise each other individually, and the one that lost in its first encounter with another, when the fight may have been more serious, quickly gives way in later disputes. However, similar behaviour is seen in large flocks of wild rooks or wood pigeons, where individual recognition seems unlikely, so probably the disputing birds come quickly to appreciate which of them would win and which would lose if the encounter were pressed.

This type of behaviour is obviously advantageous for the victor, as it gains a food item without trouble. But it is also advantageous for the loser because, given that it will lose anyway, it is most efficient for it to break off the dispute before it is hurt in order to seek for food elsewhere. It means, however, that in every such dispute the stronger and well-nourished individuals are successful, and the weak and undernourished fail. Moreover, the total numbers of the species concerned are evidently controlled by food shortage in winter, at which season disputes for food become frequent, and the weakest individuals, giving way each time, soon starve. But the naïve questioner of my opening paragraph is unaware of this part of the story. Even Lorenz seems to have been unaware of it, for though, in his book, he suggested several functions for peck-order fighting, he did not suggest that its main ecological purpose might be the food

Gulls fight over territory: the loser may fail to breed

in dispute. This may have been partly because the peck order has been studied chiefly in poultry that were given enough food by man, and not in wild birds.

Moreover, it is hard to detect that the numbers of wild birds are up against the food limit. This is an indirect result of the peck order, as first suggested by J. D. Lockie. When food is short, instead of every individual being weakened, most have enough to eat, and only the few that are lowest in the peck order have insufficient. These last die quickly and unnoticed, to be soon followed, while the food shortage continues, by those next lowest in the peck order, and so on, but at any one time the great majority are well nourished. It cannot be stressed too strongly that, in the wild, the eventual penalty for losing food disputes is starvation. Moreover, a high proportion of the population dies each year. This is so in the case of wood pigeons, for instance, among which the annual mortality amounts to about one-third of the adults and over four-fifths of the juveniles, with starvation as the basic cause (R. K. Murton, *The Wood Pigeon*, 1965). Nevertheless, ritual fighting for food items is still advantageous for the losers, because they will certainly die if they fight stronger opponents, whereas they have at least a small chance of surviving if, instead, they retreat and search for food elsewhere. So the world of birds is not, after all, idyllic. We would not enjoy a society in which one-third of our adult friends and over four-fifths of the teenagers die of starvation each year.

In territorial disputes, a trespassing neighbour not seeking to claim ground retreats immediately when threatened by the owner. When the owners of adjoining territories meet along their common boundary, they sing loudly and display vigorously, but then retire. As the usual result is, at most, a minor shift in the boundary, it is understandable that encounters of this type should be ritualised.

Much more serious and prolonged conflicts occur when a new-comer seeks to dispossess an owner of its entire territory. With robins, for instance, such struggles sometimes last for several hours or even two whole days but, nevertheless, they often consist solely of song and display. However, if one of the contestants does not give up, the two cocks may grapple with their legs and peck at each other's heads, and the bird finally defeated (by no means necessarily the newcomer) may lose many feathers and some blood, though it is not often much hurt. Why, then, should it be advantageous for the loser to give up, if a territory is essential for breeding? First, as surmised for food fighting, the combatants probably come to know which of them is likely to win if the fight were to continue; though since possession of a territory is far more important than possession of one food item, surrender may be long delayed. Even so, retreat could scarcely be advantageous unless, if the fight were pressed, the loser would be killed.

This can happen, and such birds as robins and blackbirds do at times kill each other in territorial fights. Moreover, when a robin is presented with a stuffed robin in its territory, it postures at the red breast, but when the "intruder" does not depart, it strikes hard with its beak at the base of the skull. If the loser would eventually be killed, it must be advantageous for it to break off before it is too late, and to seek for another territory elsewhere. The latter may be hard to obtain, especially late in the season, so the penalty of losing a territorial fight is a much reduced chance of leaving offspring. But there is still *some* chance of leaving offspring, and there would be none if the loser fought to a finish and were killed. Hence, while nearly all bird fighting is ritualised – which might be thought better, from the human viewpoint, than direct killing – the victor still brings disaster to the loser. And the basic cause of bird fighting is the high reproductive rate which, in a balanced population, must be equalled by the mortality.

Turning now to mankind, Lorenz and other ethologists are obviously right to claim that irrational elements come into warfare. The trumpets, the uniforms and the pomp have a deep appeal and, especially in former times, they probably helped to reduce the resistance of an opposing army in the field. A discussion on aggression was recently organised between ethologists, psychiatrists, sociologists and historians by the Institute of Biology (Symposium 13, *The Natural History of Aggression*, edited by J. D. Carthy and F. J. Ebling, 1961), and a myth has grown up that progressive scientists were there combating hidebound humanists who refused to acknowledge the animal element in human conduct (see for instance E. Ardrey, *The Territorial Imperative*, 1967, pp. 300–4). But the organisers of this symposium did not invite ecologists to contribute, so half the evidence on animal fighting – that concerned with its results – was omitted. This other half of the evidence from natural history would have supported the views of the humanists rather than the ethologists.

Of course, if Lorenz and his followers were right – that man has a strong aggressive drive that must somehow find an outlet – then the best hope of checking human warfare might be to divert or sublimate this drive into less harmful channels. But there is no evidence that human aggressive behaviour is of this nature (R. A. Hinde, *New*

Society, March 1967, p. 302). As argued by Andreski (in the Institute of Biology▪symposium just cited), the basic, though not always the immediate, cause of human warfare is the Malthusian one, that the human reproductive rate is so high that there are not enough of the necessities of life for everyone. As already mentioned, this likewise applies to birds, and the fact that, in birds, the resulting fighting is ritualised is quite secondary. Actually, the invasions of Czechoslovakia in 1938 and 1968 provide a close parallel with the ritualised fighting of birds since the aggressors obtained their ends virtually without bloodshed. But this is hardly what my liberal acquaintances want when they ask whether a fuller knowledge of bird behaviour could end wars. Even supposing that the aggressive tendencies of one nation could be diverted or eliminated by biological means, this would merely invite attack from a neighbouring state, and this deliberately for reasons of gain, not owing to some unchecked aggressive drive or instinct.

There has been serious confusion between means and ends. The bird behaviour studied by the ethologists is a means, evolved through natural selection to assist individual survival and the raising of young. Certainly it is important for men to realise that similar patterns of behaviour form an essential constituent of human nature (see N. Tinbergen, *Science*, vol. 160, p. 1411). But in man, also, they are a means, not an end. Moreover, they are not the only means. We cannot seriously hold with Ardrey (*The Territorial Imperative*, p. 5) that "we act as we do for reasons of our evolutionary past, not our cultural present". Certainly our evolutionary past is involved, but so are our culture and our reason. If we are to end human warfare, we should seek first to change not man's aggressive drive but the Malthusian compulsion. But even this would not suffice, because wars are fought not merely for the necessities of life but for riches, and under present conditions a stronger nation normally attacks a weaker one when it is likely to gain thereby. Certainly we can learn from the birds but, just as in times past the devil quoted scripture for his purpose, so now it is all too easy to draw wrong conclusions through selecting only part of the evidence from the book of Nature.

28

The urge to kill

HANS KRUUK

29 June 1972

Under certain circumstances, carnivores will kill far more than they need for food. If their prey becomes a sitting target, they seem not to know when to stop. But this is hardly a justification for the mass killings occasionally perpetrated by humans, who are aware of the consequences of their actions.

Armed with a rifle, modern man has frequently obliterated whole animal populations – sometimes for no obvious economic or other reasons. People have been known to shoot numbers of bison from a train, or to collect large heaps of antelopes in Africa, solely for the sake of a picture. Can we shrug this off as the work of psychologically abnormal individuals? Or does man kill just for pleasure, unlike any other animal species, as some conservationists claim?

While I was studying bird behaviour in a gull colony in the north of England, a total of 1449 adult gulls and many more young ones were found killed by foxes, without being eaten. In the Serengeti, in East Africa, I found on one occasion 82 Thomson's gazelle killed and 27 maimed by spotted hyaenas, again hardly utilised by the carnivores. Almost all carnivores, once they find themselves in an enclosure with suitable domestic stock, will kill and kill. In other words, man is certainly not alone among carnivorous species when he kills without reaping the benefit – we are dealing with a more general phenomenon.

Clearly, this is a phenomenon of ecological importance; clearly, also there are some interesting behavioural implications to be considered. How often does "wanton slaughter" upset the predator prey balance as much as it did in the case of man and bison? What could be the biological function of this killing for the sake of killing? Why haven't the victim species evolved better means of defence? In

How many animals died for the sake of photographs such as this one from 1898?

order to discuss these questions I will first relate some of my observations on carnivores in some detail.

During a study of predation by the spotted hyaena in the Serengeti National Park (see p. 35), it had become clear that this carnivore was a regular predator of the Thomson's gazelle, although, generally, larger species of ungulate were the preferred prey. Adult gazelle would be grabbed by solitary hunting hyaenas after a long chase, sometimes over distances as long as 5 kilometres. If the hunt was successful, the hunting hyaena and sometimes one or two others would then eat the prey entirely, leaving only stomach contents and horns. But on the occasion referred to above something completely different happened; a large number of gazelle were killed in one night by only a few hyaenas. The 82 dead gazelle were scattered over approximately 10 km², and the injured ones walked or lay between them. The latter obviously would not have survived long with their broken limbs or damaged central nervous system which caused them to walk in circles, or just to stand without moving. Nineteen hyaenas were in the area, some still eating, most of them lying down bloody-mouthed, bloated with food; they must have eaten several whole gazelles. Three out of four corpses had not been opened; of the remainder mostly only small parts had been eaten. The carcasses of which part had been eaten were scattered between the undamaged ones, suggesting that whoever killed the gazelle had interspersed his killing sequence with an occasional bite of food.

The culprits were clearly indicated by the pugmarks around and under the carcasses, apart from the fact that spotted hyaenas were the only large carnivores there. I dissected 59 of the carcasses for the inquest, and the toothmarks on the corpses and tracks on the plain made it possible to reconstruct the massacre in considerable detail. Each hyaena had quietly walked from one sleeping gazelle to the next, grabbing it randomly at any part of the body, then finishing it off with a bite in the head or neck. In the wounded animals, apparently, the head/neck bite had merely stunned the victim. Everything had taken place at a very leisurely pace, judging from the tracks, although probably the hyaenas had not stalked their prey. The night before had been unusually dark, as it was just after new moon, with very heavy rain and strong winds — a kind of weather quite rare in the Serengeti.

The scene of corpses scattered on the plain after a dark, stormy night reminded me of observations during the three breeding seasons I spent in a dune colony of black-headed gulls on the Cumberland coast. There I was interested in, among other things, predation by foxes on the gulls, and the area lent itself very well to direct observations of the foxes in the colony, and for tracking the foxes' movements on the sands at times when I could not watch it. The numbers of gulls that were killed in the colony without being eaten were staggering. Once, for example, a total of 230 adult birds were found dead, the result of one night's work by four foxes. That particular

Hyaenas occasionally catch the odd Thomson's gazelle for food; more rarely they slaughter them in large numbers. Why?

night the foxes must have grown tired of biting, as only 57 per cent of the victims had been bitten more than once as compared with 92 per cent on other nights with fewer kills. The foxes had wandered through the colonies and killed the gulls either after a short stalk and a rush, or after apparently almost accidentally stumbling on them. The number of gulls of which part had been eaten by foxes constituted less than 3 per cent of the total found dead. From the foxes' point of view, however, the gulls made up a considerable part of their diet during at least part of the year; this we concluded from analysis of the predators' faeces.

Interestingly, a small colony of sandwich terns that nested in the middle of the gull colony suffered an even higher rate of predation from the foxes. One year an estimated 12–15 per cent of the adults were killed. Of those dead terns, a relatively larger number were eaten by the killers, about 17 per cent. When evaluating the losses of adult birds, chicks and eggs to the foxes, we concluded that neither the gull nor the tern population would be able to withstand such heavy mortality over a long period. The colony would probably be doomed to extinction unless protected.

It was striking that the number of birds found dead in the colony in the early morning was clearly related to the darkness of the previous night. Gulls were significantly more vulnerable around new moon than around full moon; during heavily overcast nights with rain more were killed than during nights with clear weather. Although it was impossible to watch the foxes directly during these dark nights, it was possible to explain the dark-night slaughters after extrapolation from other observation, from tracks in the sand and by acting as a four-footed predator myself. The most striking result from those rather uncomfortable experiments was the observation that when the night was really black and wet the gulls appeared to lose the tendency to flee completely; they merely sat on their nests, and even when manually lifted off they were extremely reluctant to fly. Their direct anti-predator response seemed to vanish.

Although one is struck at first by the apparent maladaptiveness of this behaviour, the lack of response makes sense if one considers other selection pressures; any take-off under those conditions must almost certainly be followed by a crash landing. But what must be decidedly maladaptive as far as exposure to predation is concerned is the choice of that particular colony site; it allows an animal like the fox to walk leisurely from one nest to the next. This is very unusual for the black-headed gull; only a small percentage of its

H

colony sites are as accessible as the one I studied. Normally these birds nest in bogs and marshes, well out of reach of the jaws of any large land predators. Similarly, gazelle would run a serious risk of breaking legs when running over plains studded with holes, and the weather conditions under which the massacre occurred are rare.

Hyaenas and foxes are, of course, not the only carnivores which, we know, sometimes kill more than their immediate requirements demand. Almost any carnivore appears to show this kind of "surplus killing" when faced with suitable domestic stock fenced in, whether this be a fox in a henhouse, a wolf in a cattle pen or a lion in an ostrich enclosure. In that situation, as in the black-headed gull colony, normal anti-predator behaviour is absent or is prevented from occurring. There is a report of a polar bear killing 21 narwhals after they had become trapped by the ice, and lions are known to kill wildebeest one after the other when migrations of the large herds make ambushing particularly easy.

In fact, killing without eating may be considerably more common than expected, because most carnivore species have some behavioural mechanisms to deal with a superabundance of food. Caching is a very widespread phenomenon; foxes and most other members of the dog family dig holes and bury food for later use, spotted hyaenas store meat under water, leopards may drag it up into trees. In social species one individual may kill without eating and one of its companions may eat the proceeds of the hunt – this was repeatedly observed in hyaenas and in wild dogs.

It is clear, therefore, that at least in some cases there is an advantage for the predator in killing prey even when not hungry – there is possibility of utilisation at a later time, or by another member of the same social unit. It may even be that predators gain valuable experience during this "surplus killing". But against this must be weighed the disadvantages – the possibility of wiping out an important food supply which might become vital in times of stress, the waste of energy and the chance of sustaining injury while attacking a prey that can defend itself.

The unnecessary depletion of food supplies seems to be the most serious danger facing the foxes or hyaenas mentioned above. To evaluate this we will have to consider factors limiting carnivore populations. Several studies suggest that food supply does indeed play a crucial role in limiting many populations of carnivores, at least during lean seasons or years. One might expect, therefore, to find behavioural adaptations which enable these animals to make most efficient use of food resources with a minimum of waste; the observations of dozens of corpses abandoned after the kill do not fit into this picture.

To understand this apparent ecological misfit one has further to analyse the behavioural mechanism of the carnivore's foraging. Normally a hungry animal will search for some time until it contacts a potential prey; a hunt may follow which, if successful, may lead to capture and thereafter killing of the quarry. If the prey is eaten, subsequent satiation of the hunter will usually prevent further searching and hunting. While valid for carnivores, this description probably applies to hunting man as well.

If for some reason the prey species' normal anti-predator behaviour is not functioning properly, the predator may be faced with a quarry without having to go through the search and hunt stage. This may happen when the prey is penned in, or when it has to remain still in order not to damage itself in the dark, or when the predator employs long-range weapons. Our present evidence indicates that that is when the carnivore's satiation has little effect – in other words, satiation may inhibit searching and hunting, but not the capturing and killing part of the chain of foraging behaviour patterns. This organisation of behaviour may work efficiently in the day-to-day life of the predator; it prevents the animal wasting energy on foraging when it is fully fed, but also it enables the carnivore to make use of any easy opportunities that happen to come its way.

Only on the very rare occasion when suddenly a large number of prey animals becomes easily available the system misfires and food is subsequently "wasted". Wastage could be prevented if the behavioural mechanism included an inhibition of killing not merely after satiation, but after the receipt of some kind of information indicating that further carcasses could not be utilised. This may be too much to ask of carnivore evolution; it should not be too much to ask from man.

29

Sociobiology:
a new basis for human nature

EDWARD O. WILSON

13 May 1976

Sociobiology aims to offer a synthesis of evolutionary studies, genetics, population biology, ecology, animal behaviour, psychology and anthropology, using analogy with animal societies to come to some conclusions about human society.

Sociobiology is the systematic study of all forms of social behaviour, in both animals and humans. For many important reasons we may be particularly preoccupied with understanding human behaviour, and to achieve that goal requires us to pay attention to our evolutionary history, both in the recent period as hominids (during the past 10 million years) and as part of the animal kingdom as a whole. Currently the study of human behaviour is the domain of the

Edward O. Wilson by Lawrence Mynott

sociologists. They are attempting to explain our behaviour primarily by the empirical description of behaviour patterns and without reference to evolutionary explanations in any true genetic sense. The role of sociobiology with reference to human beings, then, is to place the social sciences within a biological framework, a framework constructed from a synthesis of evolutionary studies, genetics, population biology, ecology, animal behaviour, psychology and anthropology.

While being aware of the possible dangers of analogy, sociobiology puts heavy emphasis on the comparison of societies of different kinds of animals and of man. The aim is to construct and test theories about the underlying hereditary basis of social behaviour. Sociobiologists are attempting to discover the way in which the rich arrays of social organisation devised by the animal kingdom adapt the particular species to specific environmental niches. Turning more directly to man, I believe we can reject two extreme interpretations of man's behaviour proposed in recent years. We are not, as Konrad Lorenz would have us believe, at the mercy of an aggressive instinct which must be relieved periodically either through war or football matches. Certainly, we are an aggressive species, but that behaviour is finely adjusted to circumstances and capable of remaining dormant for long periods in the correct environment.

At the other extreme is the behaviourist school, exemplified by B. F. Skinner, which postulates that we are mere stimulus–response machines moulded by reward, punishment, and a few basic learning rules. That is also wrong. The truth is much more complicated than either of these two alternatives. Human behaviour must fall somewhere in between, and finding out just where is what sociobiology is all about. I should like to illustrate this by discussing a number of behavioural activities in man and animals, starting from the basis of genetic evolution.

Natural selection is the key element in evolution that determines that certain genes are transmitted more favourably from one generation to the next. The forces in the environment that exert the selection pressures operate on the manifestation of those genes. For instance, a genetically determined increase in the efficiency of reproduction or in techniques of food gathering means that the individual having the genes will produce more offspring to carry the parental genetic endowment into the next generation. Through natural selection, therefore, any device which helps to transmit a higher proportion of certain genes into subsequent generations will come to characterise the species.

For the most part, species characteristics involve physical or behavioural properties which serve to increase the chances of each individual passing on its genes to the next generation. Fitness in Darwinian terms may therefore be viewed as a particular individual's success in achieving this goal. However, with the emergence of complex social behaviour – a manifestation of the genes' more sophisticated techniques for replicating themselves – selfish behaviour becomes tempered by altruism, a form of activity that develops to exaggerated degrees in some species. This brings us to a central theoretical problem in sociobiology: how can altruism, which by its nature reduces individual fitness, possibly evolve by natural selection? The answer is kinship, the sharing of common genes by related individuals.

In a group of individuals it is quite possible that an altruistic act by a group member will increase the chances of survival or reproductive efficiency of other members, thus raising the Darwinian fitness of the population as a whole. If the group members are related genetically, it follows that an act of altruism by an individual will help to favour the transmission of its (shared) genes to subsequent generations. Natural selection will therefore select favourably for such altruistic acts and, thus, for the genes that determine them. This has been described as group selection but is more accurately termed kin selection.

The animal kingdom abounds with examples of altruistic behaviours that are instantly understandable in human terms. For instance, certain small birds, such as robins, thrushes and titmice, warn others of the approaching threat from a hawk. They crouch low and produce a distinctive thin reedy whistle. Because of its acoustic properties, the source of the whistle is very difficult to locate. Nevertheless, by giving the warning signal an individual is drawing attention to itself in a dangerous situation and a more selfish act would be to keep quiet. Dolphins will often group round an injured individual to push it to the surface where it can breath, rather than abandoning it. In African wild dogs, the most social of all carnivorous mammals, one sees altruism in a social context. When there are young in the pack most adults go off on a hunting expedition leaving the pups to be cared for by an adult, usually but not always the mother. When the hunters return they regurgitate food for all the animals in the camp, which occasionally includes sick and crippled individuals too.

Chimpanzees, man's closest relative, display an interesting form of altruism when they temporarily abandon their normally vege-

African wild dogs, the most social of all carnivores, live and hunt in packs and display a high degree of sharing and altruistic behaviour

tarian diet and indulge in meat eating. Adult chimps – usually the males – sometimes hunt and catch young monkeys, and through a system of elaborate begging gestures other members of the troop can share in the catch. Curiously, chimps do not share in this way when they are eating leaves and fruit.

We have to look to the social insects, however, to encounter altruistic suicide comparable with that sometimes displayed by man. A large percentage of ants, bees and wasps are ready to defend their nests with insane charges against intruders. Such attacks may involve inevitable suicide through heads being ripped off (as in the social stingless bees of the tropics), viscera torn out (honey-bee workers), or the whole body being blasted by "exploding" glands (an African termite). In all these cases the suicidal deterrent is inflicted by individuals who are sterile or have low reproductive potential. But by their sacrifice they are (in terms of Darwinian fitness) increasing the reproductive chances of their fertile relatives, thus ensuring that their (shared) genes are transmitted to future generations.

It is the unusual distribution of reproductive potential in the social insects that allows the emergence of exaggerated biological altruism. But, as we have seen, altruism appears to have been selected evolutionarily in higher animals too, mediated by kin selection. What can we say of man? If we look back into our immediate evolutionary past we see that almost certainly the social unit was the immediate family and a tight network of close relatives. Such social cohesion, combined with a detailed awareness of kinship made possible by high intelligence, is certainly very favourable for the operation of kin selection and may explain why this evolutionary force is stronger in humans than in monkeys and other animals.

An essential change of gear in the emergence of man, of course, was when cultural evolution became more important than biological evolution, a change which occurred perhaps about 100 000 years ago. As a result it seems clear that human social evolution is more cultural than genetic. Nevertheless, I consider that the underlying emotion of altruism, expressed powerfully in virtually all human societies, is the consequence of genetic endowment. The sociobiological hypothesis does not, therefore, account for differences between societies but it could explain why human beings differ from other mammals and why, in one narrow aspect, we more closely resemble social insects. It is salutary to consider the possibility that, with the extreme family dispersal characteristic of advanced industrialised society, altruistic behaviours will decline through the loss of group selection, a process that could spread over perhaps two or three centuries.

On the opposite side of the coin to altruism is aggression, one of the most important and widespread organising techniques in the animal kingdom. Animals use it to stake out their territories and to establish and maintain their group hierarchies. Some people argue that humans share a general aggressive instinct with animals and that it must be relieved, if only through competitive sport. But if we look closely at a number of species we see that aggression occurs in a myriad of forms and is subject to rapid evolution – there is no general instinct. For instance, we commonly find one species of bird or mammal to be highly territorial, employing elaborate, aggressive displays and attacks, while a second, closely related, species shows no territorial behaviour. If aggression were a deeply rooted instinct, such differences would not arise.

The key to aggression is the environment. We see that, despite the fact that many kinds of animals are capable of a rich, graduated repertory of aggressive actions, and despite the fact that aggression is important in their social organisation, it is possible for individuals to go through a normal life, rearing offspring, with nothing more than occasional bouts of play fighting and exchanges of lesser hostile displays. Aggression may increase under conditions of social stress, the result perhaps of crowding or limitations in food supplies. We can only conclude that the evidence from comparative studies of animal behaviour cannot be used to justify extreme aggression, bloody drama, or violent competitive sports practised by man.

This brings us to a crucial issue with which sociobiology has to grapple: what are the relative contributions to human behaviour of genetic endowment and environmental experience? It seems to me that we are dealing with a genetically inherited array of possibilities, some of which are shared with other animals, some not, which are then expressed to different degrees depending on environment. Our overall social behaviour most closely resembles that of the species of Old World monkeys and apes, which, on the basis of anatomy and biochemistry, are our closest relatives. This is just what one would expect if behaviour is not based on experience alone but is the result of interplay between experience and the pattern of genetic possibilities. It is the evolution of this pattern that sociobiology seeks to analyse.

For at least a million years – probably more – man engaged in a hunting–gathering way of life, giving up the practice a mere 10 000 years ago. We can be sure that our innate social responses have been fashioned largely through this lifestyle. With caution, we can, therefore look at the dwindling number of contemporary hunter–gatherers and hope to learn something about our basic social organisation. We can compare the most widespread hunter–gatherer qualities with similar behaviour displayed by some of the non-human primates that are closely related to man. Where the same pattern of traits occurs in man – and in most or all of those primates – we can conclude that it has been subject to little evolution. Variability in traits implies evolutionary plasticity.

The list of human patterns that emerges from this screening technique is intriguing:

(1) The number of intimate group members is variable, but is normally 100 or less.

(2) Some degree of aggressive territorial behaviour is basic, but its intensity is graduated and its particular forms cannot be predicted from one culture to the next.

(3) Adult males are more aggressive and are dominant over females.

(4) The societies are largely organised around prolonged maternal

care and extended relationships between mothers and children.

(5) Play, including at least mild forms of contest and mock aggression, is keenly pursued and probably essential to normal development.

In addition to this list are a number of unique human characteristics, so distinct that they can be safely classified as genetically based: the overwhelming drive to develop some form of true semantic language, the rigid avoidance of incest by taboo, and the weaker but still strong tendency for sexual division of labour. That this division of labour persists from hunter–gatherers through to agricultural and industrial societies is highly suggestive of a genetic origin. We do not know when this trait emerged in human evolution, nor how resistant it is to the continuing and justified pressures for women's rights.

At this point I should stress a constant danger in sociobiology, and that is the trap of the naturalistic fallacy of ethics which uncritically concludes that what is should be. The "what is" in human nature is the legacy of a long heritage as hunter–gatherers. Even when we can identify genetically determined behaviour, it cannot be used to justify a continuing practice in present and future societies. As we live in a radically new and changing environment of our own making, such a practice would invite disaster. For example, the tendency under certain conditions to indulge in warfare against competing groups may well be in our genes, having been advantageous to our Neolithic ancestors but it would be global suicide now. The drive to rear as many healthy children as possible, once the path to security, is now environmental disaster.

Sociobiology can help us understand the basics of human behaviour and the fundamental rules that govern our potential. We will need to know how, genetically, certain types of behaviour are linked to others. We must understand the mechanism and the history of the human mind. The special insights made possible by sociobiology can join with the social sciences to create a new study of man, one by which we might hope to steer our species safely in the difficult journey ahead.

30

The new synthesis is an old story
SCIENCE AS IDEOLOGY GROUP OF THE BRITISH SOCIETY FOR SOCIAL RESPONSIBILITY IN SCIENCE
13 May 1976

Although Wilson and his fellow sociobiologists claimed that they had been misinterpreted, their ideas attracted strong criticism for justifying the status quo in human society. This article is typical of the uproar that broke out on both sides of the Atlantic after Wilson's book was published.

The proposed new discipline of sociobiology is, in the words of its principal exponent, E. O. Wilson, "the systematic study of the biological basis of all social behaviour" which will ultimately encompass all of human history since "sociology and the other social sciences as well as the humanities are the last branches of biology waiting to be included in the Modern Synthesis" (*Sociobiology: The New Synthesis*). The theory proposes to depict all of human society

and its growth, development, present state and future prospects in biological and genetic terms.

In the United States the new discipline is appearing in college curricula and a school text has been written in which students are required to give genetic evolutionary answers to such questions as "Why do children hate spinach while adults like it?", or rather less innocuously "Why don't females compete?", "Why aren't males choosy?", and "How did the pair bond become part of human nature?" (I. deVore, G. Goethals, R. Trivors, *Exploring Human Nature*, Unit 1, Educational Development Corporation, Cambridge, Mass, 1973).

Despite disclaimers of political intent by sociobiologists and despite the similarity of sociobiology to other works of biological determinism such as Konrad Lorenz's instinctual theories and C. D. Darlington's genetic explanations of all of history (*The Evolution of Man and Society*, 1969), the media have responded quickly and enthusiastically to Wilson's book. The *New York Times* ran a front-page piece (28 May, 1975) which stated "Sociobiology carries with it the revolutionary implications that much of *man's* behaviour towards *his* fellows may be as much a product of evolution as the structure of the hand or the size of the brain" (our emphasis). A review in *New Society* (9 March, 1976), although critical hailed the text as a "truly monumental book" and a "valuable and outstanding work of scholarship".

The potential implications of the new discipline have been analysed by the Sociobiology Study Group of Science for the People in Boston. In a series of detailed criticisms (Science for the People, 16 Union Square, Somerville, Mass) they have examined the new discipline for its scientific content and rigour, its ideological assumptions and its political extrapolations. On all these aspects there is room for concern. Two major premises are embedded in its arguments about the continuity between human and animal social behaviour. The first is that *all* human societies share certain specific human behaviours which constitute a universal "human nature". The second is that these behaviours are the expression of specific genetic structures and thus are a result of evolutionary adaptation through natural selection.

The specific human behaviours presumed by Wilson to be genetically coded include aggression, allegiance, altruism, conformity, ethics, genocide, indoctrinability, love, male dominance, the mother–child bond, military discipline, parent–child conflict, the sexual division of labour, spite, territoriality, and xenophobia.

These elements are combined into a view of a presumed universal human economy based on scarcity and unequal distribution of resources and rewards (Wilson, p. 554).

One is inclined to dismiss this view of the universal human society as the vision of a .person completely bound by a near-sighted cultural chauvinism. The image of society presented depicts today's European and American capitalist societies. It ignores ethnographic documentation that contradicts this conception of social organisation. Societies exist which are *not* differentiated by role sectors, which are *not* differentiated by higher and lower strata, and which are *not* characterised by deprived access to social rewards (see M. Sahlins, *Stone Age Economics*, Aldine-Atherton, 1972). Wilson (pp. 564, 574) is aware of exceptions to his presumed universals but claims that the exceptions are "temporary" aberrations or deviations.

The major sociobiology argument, however, rests on a presumed genetic basis to human social traits and on presumed similarities between human societies and other animal societies. Specific genetic structures are postulated to exist for the social behaviours listed above. There is no direct evidence for the existence of such structures. Modern biology has not discovered any part of DNA that codes for any human behaviour, let alone for such specific traits as altruism, conformity, domination, or spite. Specific genetic structures for particular traits are thus speculations woven into the argument only by assumption.

In Wilson's book the distinction between assumption and fact is often confused. For example, on p. 554 he says, "Dahlberg (1947) showed that *if* a single gene appears that is responsible for success and upward-shift in status . . . . Furthermore there are *many* Dahlberg genes" (our emphasis). The effect of this confusion is to leave the reader with the idea that there is a firm basis for the existence of genetically coded traits while at the same time permitting Wilson and his defenders (for example, Robert May in *Nature*, 1 April, 1976) to argue that in fact they are only speculating that such genes may exist.

In the absence of direct genetic evidence, the biological links between animals and humans must be established by observing similarities between human and animal behaviour. Sociobiology, in common with a long biological tradition, uses metaphors from human societies to describe animal societies and in so doing posits behavioural similarities between humans and animals. The classic examples of this practice, which long antedate sociobiology, are the

use of the terms "slavery" and "monarchy". Human slavery involves members of one's own species, the use of force, and the use of
the slave as a commodity and as a producer of economic surplus.
"Slavery" in ants involves "slave-making" species of ants which
capture immature members of "slave" species. When the captured
ants hatch they perform housekeeping tasks with no compulsion as
if they were members of this captive species. A more apt term for this
might be "domestication" rather than "slavery". Human slavery
has nothing to do with ants except by weak and spurious analogy.
Similarly the so-called "queen" bee may be more a captive of the
"workers" than their "ruler", since in many species she is only a
laying machine bloated with eggs, forced by "workers" to remain in
one place and to reproduce continually. Sociobiology uses
metaphors from human social arrangements to find culture
(Wilson, pp. 173, 559), division of labour (p. 299), aesthetics
(p. 564), and role playing (p. 290) among animal societies. From
there it is a short step to assert that magic, religion, ritual and
tribalism are evolutionary genetic adaptations in human societies
(p. 560). Human institutions thus appear natural, universal and
genetically based.

Two further technical points are worth discussing. The first concerns specific tests of a genetic model of human cultural evolution.
Population genetics is capable of making specific *quantitative* predictions about rates of change of characters in time and about the
degree of differentiation between populations. In addition there
exist hard data on genetic differentiation between populations for
biochemical traits. Both the theoretical allowable rates of *genetic*
change in time and the observed *genetic* differentiation between
populations are too small to agree with the very rapid changes that
have occurred in human cultures historically and the very large
cultural differences observed between contemporaneous populations. Sociobiologists acknowledge this problem. But rather than
evaluate the theory in hard terms, the problem is evaded through the
introduction of a fudge factor, the "multiplier effect" (a phrase
borrowed from Keynesian economics). The multiplier effect postulates that very small differences (Wilson, pp. 11 and 572) in
genotypic frequencies can result in major cultural differences. To
account for the fact that non-human animal societies do not show
equally rapid evolution and equally dramatic interpopulation variation in social traits, an additional effect, the threshold effect, is
postulated. It argues that organisms must reach a certain (unspecified) level of social complexity before the multiplier effect will

operate (p. 573). They serve to seal off the theory from tests against the real world of cultural change and diversity.

The second technical point concerns the question of altruism. A significant accomplishment of sociobiology, for the naturalist, has been to generate a coherent solution (see Wilson's chapters 2–26) to the problem posed by the existence of self-sacrificing behaviours, bearing a superficial resemblance to human altruism, in certain species. Unfortunately, the word "altruism" has been appended to this, by a process similar to the use of words like "slavery". This animal "altruism" poses a problem for the neo-Darwinian theory of evolution, as it did for Darwin, because the theory is based absolutely upon competition between individual sets of genes (within individuals of a species) to drive evolution.

How then can altruism have evolved? Theoretical work of Wilson and others has shown how certain breeding systems could support such organic evolution. Wilson puts forward the supposition that human altruism, too, has a comparable genetic basis. Seen thus, it would appear to be the resultant of a blind process of competition between individuals for *genetic* success, throughout prehistory.

Sociobiology represents yet another attempt to employ the methods and perspective of biology to deal with the problems of human society. This practice has had a long and frequently ugly history. Throughout the development of modern science, ideologues have dreamed of developing theories in the name of science which could legitimise the massive social inequity generated by Western capitalism.

Eight years before the publication of the *Origin of Species*, Herbert Spencer, the English social theorist, argued that the elimination of poverty was unnatural since "the poverty of the incapable, the distresses that come upon the imprudent, the starvations of the idle . . . are the decrees of a large, far-seeing benevolence . . . under the natural order of things society is constantly excreting its unhealthy, imbecile, slow, vacillating, faithless members" (*Social Statics*, p. 353).

Sociobiology is the modern successor to Spencer's social Darwinism. Survival of the fittest is replaced by the theoretical framework of population genetics with the necessary postulates and fudge factors added to make the theory an untestable whole. Its ideological basis is clear. Yet in a society where there is a belief that science is a rational, objective and value-free activity, there may be resistance to the idea that science or scientific theories have ideological content.

But there is no justification for treating scientific theories differently from others, or scientists as experts to be deferred to for social and political guidance. In order to uncover this ideological content, the theories must be located and analysed against the contemporary social situation in which they are formulated. Sociobiology arrives at a time when wide ranging challenges to the existing social order are being made. Unlike the more particularistic theories (such as race and IQ), it avoids specific charges of racism and sexism because its all-encompassing nature enables it to explain the foci of all these theories at once. It is, of course, racist and sexist – and classist, imperialist and authoritarian, too. By assuming that there are specific genetic structures which embody particular manifestations of human behaviour, sociobiology neatly defuses these challenges by exposing their supposed biological hopelessness. Whether or not Wilson likes it or intends it, his work is part of this. *Sociobiology: The New Synthesis* will be used and, indeed, is already being used to justify the biological inevitability of the status quo. The trouble with sociobiology is not merely that it makes assumptions and extrapolations that are misleading. More importantly, it directs our gaze to evolutionary theory as the key to the limits of human nature, and it implies in its long agenda for scientific research that a better social order must await the findings of scientists.

What is needed, and what sociobiology clearly cannot provide, is a clear understanding of the uniqueness of the human species. We are unique animals, who, while having a physical basis in organic life, transcend it. We alone possess the capacity for language, culture and the ability consciously to transform our environment. It is to these capacities that the creative roots of a theory of our history and our social problems must look. The development of such autonomous cultural and social theory, taking as its basis those qualities unique to human life, is currently stultified in the Anglo-American world by a deep-rooted deference on the part of many social scientists towards biological knowledge. This deferential posture ensures a readiness of the part of the human sciences, and hence the community in general, to adopt biologically moulded models into their own thought. It is maintained by the primacy of biology in school-level science education. But one is taught early on how organic life is distinguished from its component molecules by various emergent qualities and capacities, whilst recognising their material continuity with inorganic nature.

We would like to see those other emergent qualities, that distinguish humanity equally profoundly from its organic material basis, treated also as basic in our educational curricula. Genetics has as little to tell us about human societies as nuclear physics has to tell us about genetics. In the same way that we do not turn to physics in order to understand genetics, we should not turn to genetics in order to understand human history and culture.

31

Biological limits to morality

ROGER LEWIN

15 December 1977

Do our concepts of right and wrong exist only as a byproduct of our ability to reason? Or can their existence be explained with reference to evolutionary biology? In 1977 a conference took place in Berlin in an attempt to resolve these issues.

The late Conrad Waddington wrote a book called *The Ethical Animal*. He was of course referring to *Homo sapiens*, the only animal whose life is governed by a series of overtly specified rules, norms and values. In the main, individuals interact with others according to what they accept as being either "good" or "bad" and this, as every cultural anthropologist is quick to point out, varies not only between different societies but also within a single society. These kinds of social norms – which may be gathered under the collective term, ethics – have been within the province of philosophical dispute for two millenia, ever since Aristotle and Plato set up roughly opposing views on their origins. Aristotle saw moral behaviour as the result of *de novo* inventions of human mind, whereas Plato took a more naturalistic approach, seeing in ethical norms the expression of a more basic biological strategy for survival. It was the status of this philosophical divide that was the concern of participants of a Dahlem conference held recently in West Berlin. More than just philosophers took part: psychologists and biologists joined in the fray too.

The conference had set itself the task of deciding just how much of the concept and practice of morality could be explained by the naturalistic approach: what does biology tell us about ourselves as ethical animals?

The naturalistic school of human behaviour was boosted considerably when Charles Darwin published his *On the Origin of Species* in

1859, an event that revolutionised man's view of his place in nature: we were forced to accept that we are a part of nature rather than apart from it. The Platonic position was further strengthened, first, by the rise of ethology during the past half century and, secondly, by the recent emergence of "sociobiology", a term coined by Harvard biologist Edward Wilson in his controversial book of the same name published in 1975 (see pp. 213 and 219). Sociobiology purports to offer explanations of animal behaviour, including human, in terms of survival strategies exploited by individuals within species. As things worked out at the Dahlem meeting, the term "naturalistic approach" in the conference task was effectively replaced by "sociobiology". The question, therefore, became, Are moral norms simply part of a clever strategem for getting us to behave in a way dictated by our genes?

In the arena of natural selection, a species survives according to how well it is suited to the prevailing environment, suitability being measured in terms of both physical characteristics and behavioural repertoire. A species well adapted to its environment may thrive, whereas one that finds it difficult to make a living may become extinct, or change. In the unremitting confrontation between a species and its environment, it is not the animals *as a group* upon which selective pressures act, it is the *individual* animal that is the so-called unit of selection. For instance, if an individual has a heritable characteristic that gives it a competitive edge over its fellows, whether in access for food or for mates, then that individual may produce more than the average number of offspring. As time goes on that heritable trait will become more common as the growing number of descendants thrive. The outcome of this is that one should expect animals to pursue essentially selfish patterns of behaviour because those that did not in the past did not survive.

As Richard Dawkins illustrated so clearly in his book *The Selfish Gene*, such selfishness may come in the guise of helping others. In performing so-called altruistic acts – such as giving an alarm call when a predator threatens the group, or helping another to have access to a female – an animal is in fact helping himself because the recipients of his actions may be his relatives (they share his genes). By being altruistic the individual is ensuring the survival of his genes, not necessarily through himself but through another. Such altruism can arise only in animals that live in social groups populated by closely-related individuals.

Self-centred altruism of the sociobiological type can, however, develop among non-kin in more complex social systems, groups of animals that live together for a long period and have sufficient brain power for individual recognition and memory. (Brain power may not in fact be required absolutely for this more developed form of altruism, for instance in certain kinds of fishes, but while we are contemplating the issue of morality it is reasonable to include it.) If an individual gives a helping hand to another "knowing" that in the long run the favour will be returned, then so-called reciprocal altruism is possible. Helping others is beneficial to oneself as long as there is the certainty that help will be forthcoming from one's former beneficiaries when it is needed. The whole system works on a fine balance sheet of "trust", and in an intelligent animal it could be backed up with the psychological tools of sympathy and gratitude, moral indignation and shame, and other emotional responses.

These refinements of reciprocal altruism begin to sound very familiar, for it is precisely what we experience as humans. (It also has distinct overtones of some aspects of morality.) In human prehistory, our ancestors' social life provided a fertile soil for the germination of reciprocal altruism, probably the best example in the whole of the animal world. About three million years ago primitive hominids began to develop a subsistence *economy* as opposed to a more basic system of individualistic opportunism. Instead of feeding individually, albeit within the context of a strongly developed social group (like modern chimpanzees, for instance), they began to elaborate a mixture of hunting and gathering, the spoils of which were taken back to a home base to be shared out.

The social demands of operating this kind of food-sharing economy must have been enormous, in terms of distribution of tasks, psychological restraint and cooperation with others. All primate social groups display a rich network of individual interactions, but the economic content of our ancestors' newly developed way of making a living escalated social life to heights unscaled by any other animal. It was a way of life built on richly elaborated reciprocity, a theme that runs counter to the inescapable genetic competition between individuals. Ancient hominids pursued a hunting and gathering way of life for virtually three million years, beginning its replacement by agriculture a mere 10 000 years ago. That is a very long period of time during which the social and psychological demands of a particular lifestyle can become stamped into the intellectual machinery.

The enormous social complexity of the hunting and gathering economy would eventually have made it inoperable without a system of social rules, rules that may be described as social norms or even a morality. Guidelines for social conduct to any degree of utility would of course have been impossible without some form of spoken language in which to elaborate them. By at least half a million years ago – the time at which the basic grade of *Homo sapiens* was beginning to emerge from its predecessor, *Homo erectus* – the human animal was already a highly intelligent and in many ways potentially unpredictable creature. Accommodated within a set of social norms, human unpredictability would have been more manageable, producing a more stable social environment.

At this point we readily recognise the basis of modern *Homo sapiens*, a rule-making and rule-following creature. Against this background one can construct a number of hypotheses to account for the elaboration and adherence to social rules in human societies. The strongest from the sociobiological viewpoint is that the greater part – if not all – of human behaviour, both in structure and content, is determined by genetic imperative. In this case human morality is seen as the morality of the gene, a basic biological strategy for individual survival. One may *feel* intuitively that this must be wrong, but if indeed it were correct, such a feeling would not be inconsistent with it being so: we could be fooled by our genes because the system works better that way!

A second hypothesis, also sociobiologically based, but slightly weaker, is that humans have a specific predisposition for imbibing social norms. Psycholinguists of the Chomsky school argue that humans have a specific language-acquisition device; otherwise, they say, it is inconceivable that children would be able to learn the grammatical structure of their native tongue from the chaotic input they receive from the world around them. This is a conceptually attractive idea (currently, however, it is under attack) and it provides a useful analogy for the second thesis. If an important degree of social conformity has in the past been central to the success of human groups, then it would clearly be advantageous if individuals were able readily to "recognise" and assimilate those factors in the environment that underpin group morality. In this case the structure, but not the content, of moral behaviour can be seen as the consequence of genetic heritage. By contrast with the stronger thesis, this second notion would allow much more flexibility in social norms and we would avoid being stuck with a piece of behaviour that in the past might have been useful but which now is not only redundant but also possibly antisocial.

The third position on the origins of morality – and the one that rejects sociobiology totally – is that evolutionary forces moulded a human brain of high intellectual capacity, and one product of this intellect is a system of artificially created rules. Here, neither the structure nor the content of morality is impinged upon by genetic imperative.

The problem, of course, is how one decides between them. This is more than a matter of mere intellectual curiosity, because if, for example, the first thesis turned out to be correct, then we might have to accept that there are important limits to the way that society can be ordered as a matter of will, a position that is less than attractive to the Marxist school.

As with all great scientific questions, there is no *single* test that will settle the issue one way or the other. The problem is made more difficult because even the strong sociobiological position does not demand that *all* human moral behaviour must be uniform: it clearly is not in any case. It is quite possible that differences between societies are genetically based but one would have to postulate very keen selection pressures for this to occur. Faced with the impossibility of examining social variation from a sociobiological perspective (for the moment anyhow), one is forced to look for universals. Once again, if one *does* find important behaviour patterns that are more or less universal to all societies, this does not necessarily mean that they are genetically based – but it would make one very curious about them.

Possible candidates for "moral universals" are the capacity for empathy, inhibition against killing one's fellows, incest prohibition, social conformity, "sex roles", and the nuclear family. Of these, incest prohibition appears to be the most accessible to analysis.

Despite cries of protest from cultural anthropologists, there does appear to be a universal prohibition against copulation with one's closest kin. An interesting point here is that animals that live in social groups where incest is possible all appear to have mechanisms that effectively prevent incest. The principal argument suggesting that incest is harmful is genetic, but there is disagreement about whether it is because of potential damage to the offspring or because of the benefits of outbreeding as against inbreeding. In any event, the result is the same. There are psychoanalytic and economic explanations of incest prohibition in human society too. But the parallel between the functional *biological* explanations in animals and humans is persuasive of a point for sociobiology.

The argument for psychological differentiation between men and women, though weaker than the incest case, is still persuasive in a general way. The question is, how immutably are such systems currently embedded in society? What would happen if, for instance, all rules against incest were removed and men and women were encouraged to share their roles equally? It is unlikely that the nation would indulge in an orgy of mass incest, because there appear to be very real and powerful mechanisms preventing sexual attraction between individuals that have been brought up together. The case of sex roles is undoubtedly more flexible but there are many people – and not just male chauvinists – who would not be surprised if, even given the opportunity in the right environment to do otherwise, many women would incline towards child rearing rather than committing themselves to direct economic activity.

An important meeting point for investigations into the nature and origins of morality is the growing child. Traditionally, philosophers have focused on the beliefs and behaviour of adults in their attempts to understand and define morality. But by examining the way that children come to articulate and follow moral norms one may be able to achieve a more empirical analysis of morality. There are, of course, pitfalls, just as there are with all excursions into developmental psychology: the investigator may be measuring what he sees as the essentials of morality, but he will bring to the definition not only his own personal prejudices but also those of the philosophical approach of the West. (Incidentally, each of the three groups at the Dahlem meeting – biologists, psychologists and philosophers – drew up definitions of morality; there were *some* points in common between them, but the area of disagreement was impressive.) In launching themselves into this area of investigation researchers are clearly exposing themselves to more than the usual degree of intellectual hazard. Care is needed.

If the very strongest sociobiological thesis were to be correct, then one may question the use of the term morality at all because it is usually used in the context of the possibility of behaving other than morally. An unrelenting genetic imperative would remove that element of choice. There are, however, very few reputable sociobiologists who champion the view that we are mere automatons marching to the orders of our genes. Indeed, there are also few who would commit the naturalistic fallacy of suggesting that what is natural is right: *is* does not imply *ought*. It would be foolish, however, to reject for ideological reasons any possibility of there being an *is*, for this would abrogate a vital part of our humanity: intellectual curiosity. At the same time, one cannot ignore the political content of the whole issue: awareness, after all, is a powerful weapon against political misuse.

PART SIX

What Makes Animals Different from Us?

Homo sapiens may be defined as "the thinking primate", but what do we mean by thinking? After all, not all forms of logical reasoning are beyond even rats and pigeons (see Part I). Like it or not, we are going to have to draw the line between ourselves and the rest a little more carefully (p. 235). If we have evolved our mental abilities, their roots must lie in our subhuman ancestors.

For most biologists, consciousness is a uniquely human attribute; there is no evidence that other species possess our capacity for self-awareness. Human consciousness expresses itself in language, that gives us the power to look into our own minds. Despite the painstaking efforts of primate researchers, there is still no evidence that animals have the capacity to learn and use language. Chimpanzees can certainly learn and use *symbols*, which is in itself a considerable achievement for an animal; they can even string symbols together in order to communicate with each other or a human companion. But they do not use the symbols in a truly linguistic fashion.

It is perhaps a little unfair on animals to label them dumb because they cannot do some of the things that we can. Instead, why not look at the abilities they do possess? It is more interesting to ask of the signing chimp not "Is it language?" but "Why can they do it?". Chimpanzees in their natural state do not use an extensive sign language but they do manage to communicate remarkably complex information to each other, apparently with no more than a meaning look or a gesture (p. 244). Perhaps the tables should be turned, and latter-day Dr Doolittles attempt to learn the "language" of chimps!

We somehow expect intelligent behaviour from our close relatives the apes. It comes as more of a surprise to find that those denizens of the Skinner box, the rat and the pigeon, whose intellect was held in such low esteem for so long, also have some claim to intelligence, though of a simple kind (p. 261). It is even more startling to find that most robot-like of creatures, the honey bee, showing signs of what looks remarkably like the workings of a mind (p. 272). Evolution seems to have allowed even simple creatures just enough flexibility in their wiring to cope with the chance happenings of their everyday lives. Can humans claim any more?

The latest effort to give animals credit for greater mental capacity than before is more than a return to 19th century anthropomorphism. Animals are not machines but they are not people either. We should not attribute to them our own subjective experiences of joy or pain. But if we accept that they do have subjective experiences, we must make every effort to understand them. This is at the heart of research into the welfare of animals exploited by humans, in laboratories or on farms. Treatments that cause unnecessary suffering are forbidden by law but what constitutes suffering is still a matter for debate (p. 298).

The scientific community engages in a certain amount of double-think where anthropomorphism is concerned. Officially it is anathema but in unguarded moments almost any researcher will discuss his or her animals in affectionately human terms. This tendency has come into the open in the publication of several popular accounts of primate research (p. 304). The general public lacks the cool objectivity of the scientific community; it wants its emotions to be engaged when it reads about the antics of its near biological relatives. Are scientists letting the side down when they satisfy this demand? Or are they just being honest?

32

Consciousness – a Just-So story

NICHOLAS HUMPHREY
19 August 1982

At some point in evolution, our prehuman ancestors became conscious. Why was consciousness worth having? And assuming that no other animal species is conscious (perhaps not a universally acceptable assumption), why has none of them followed the same evolutionary route?

Biologists who have thought, but not thought enough, about consciousness will be found toying with two contradictory ideas. First – the legacy of the positivist tradition in philosophy – that consciousness is an essentially private thing, which enriches the spirit but makes no material difference to the flesh, and whose existence either in man or other animals cannot in principle be confirmed by the objective tools of science. Second – the legacy of evolutionary biology – that consciousness is an adaptive trait, which has evolved by natural selection because it confers some (as yet unspecified) advantage on those individuals who possess it.

Put in this way, the contradiction is apparent. Biological advantage means an increased ability to stay alive and reproduce; it exists, if it exists at all, in the public domain. Anything which confers this kind of advantage – still more, anything whose evolution has specifically depended on it – cannot, therefore, remain wholly private. If consciousness *is* wholly private it cannot have evolved. Or if it has evolved, it must in Hamlet's words be but private north-northwest; when the wind is southerly it must be having public consequences. If the blind forces of natural selection have been able in the past to get a purchase on these consequences, so now should a far-seeing science.

Yet scholars will, I suspect, continue to tolerate the contradiction, paying lip service both to the privacy and to the evolutionary adaptiveness of consciousness, until they are offered a plausible account of just wherein the biological advantage lies. At present, so far from having a testable hypothesis which we could apply to species other than our own, we lack even the bones of a good story about consciousness in human beings. I here present one: a Just-So story.

But first some pointers to what, in the context of this story, I take "consciousness" to mean. I rely on there already being between us the basis for a common understanding. I assume that you yourself are another conscious human being; that you have a personal conception of what consciousness is like; that you have experienced, waking and sleeping, both its presence and its absence; and that having noticed the contrasts you have already formed some notion of what consciousness is for. I assume, moreover, that although you may never have had occasion to pronounce on it, you will not find it difficult to recognise someone else's pronouncements (mine, below) as true to your own case.

Provided, that is, you are in fact a conscious human being and not, as it happens, an unconscious robot or a philosopher from Mars. Provided, also, that you have not been too much influenced by Ludwig Wittgenstein. When Wittgenstein in his *Philosophical Investigations* alluded to consciousness as a "beetle" in a box – "No one can look into anyone else's box, and everyone says he knows what a beetle is only by looking at *his* beetle . . . . Everyone might have something different in his box . . . . The box might even be empty" – he chose the name of a thing which has no obvious use to us, and thereby implicitly ruled out the possibility that the things in our several boxes might bear a functional resemblance to each other. But suppose the thing in the box had been called, let's say, a "pair of scissors". One person's pair of scissors might indeed look rather different from another's: long scissors, short scissors, scissors made of brass or steel. But scissors, to be scissors, have to cut. There is really no danger that what we both agree to call a "pair of scissors" could in my case be a jelly-baby while in your case is empty air.

From all I know about myself, what strikes me – and seems to give some kind of cutting edge to consciousness – is this. The behaviour of human beings, myself included, is in every case under the control of an internal nervous mechanism. This mechanism is responsive to and engaged with the external environment but at the same time operates in many ways autonomously, collating information, hatching plans, and making decisions between one course of action and another. Being internal and autonomous, it also, for the most part, operates away from other people's view. You cannot see directly into my mechanism, and I cannot see directly into yours. Yet, *in so far as I am conscious*, I can see as if with an inner eye into my own.

During most of my waking life I have been aware that my own behaviour is accompanied by certain conscious feelings – sensations, moods, desires, volitions and so on – which together form the structure and content of my conscious mind. So regular indeed is this accompaniment, so rarely does anything happen to me without its being either preceded or paralleled by the experience of a conscious feeling, that I have long ago come to regard my conscious mind as the very same thing as the internal mechanism which controls my bodily behaviour. If I ask myself *why* I am doing something, like as not my answer will be framed in conscious mental terms: I am doing it *because* I am aware of this or that going on inside me. "Why am I looking in the larder? Because I'm feeling hungry . . . . Why am I raising my right arm? Because I wish to . . . . Why am I sniffing this rose? Because I like its smell."

Thus consciousness (some would say "self-consciousness", though what other kind of consciousness there is I do not know) provides me with an explanatory model, a way of making sense of my behaviour in terms of which I could in no way otherwise devise. To the extent that it succeeds, it does so presumably because the workings of my conscious mind do in reality bear some kind of formal (if limited) correspondence to the workings of my brain. "Hunger" corresponds to a state of my brain; "wishing" corresponds to a state of my brain; even the organising principle of consciousness, my concept of my "self", corresponds to an organising principle of brain states. Not that physiologists have yet come up with an analysis of brain activity along these lines. But that, for the moment, is their problem, not mine. As a child of the evolutionary process, whose ancestors have been in this business for many millions of years, I am, in relation to my own behaviour like the ancient astronomer in the picture, who has found a way of looking in directly on the wheels and cogs which move the stars across the heavens: the stars are my behaviour, the cogwheels are the mechanism that controls it, and the astronomer peering in on them is I myself.

So what? So, once upon a time there were animals ancestral to man who were not conscious. That is not to say that these animals lacked brains. They were no doubt percipient, intelligent, complexly motivated creatures, whose internal control mechanisms were in many respects the equals of our own. But it is to say that they had no way of looking in upon the mechanism. They had clever brains, but blank minds. Their brains would receive and process information from their sense organs without their minds being conscious of any accompanying sensation; their brains would be moved by, say, hunger or fear without their minds being conscious of any accompanying emotion; their brains would undertake voluntary actions without their minds being conscious of any accompanying volition. . . . And so these ancestral animals went about their lives, deeply ignorant of an inner explanation for their own behaviour.

To our way of thinking such ignorance has to be strange. We have experienced so often the connection between conscious feelings and behaviour, grown so used to the notion that our feelings are the causes of our actions, that it is hard to imagine that in the absence of feelings behaviour could carry on at all. It is true that in rare cases human beings may show a quite unexpected competence to do things without being conscious of their inner reasons: the case, for example, of "blind sight", where a patient with damaged visual centres in the brain can point to a light without being conscious of any sensation accompanying his seeing (and without, as he says, knowing how he does it). But the patient himself in such a case confesses himself baffled and you and I will not pretend that that would not be our reaction too.

Such bafflement, however, was one among the many things our unconscious ancestors were spared. Having never in their lives known inner reasons for their actions, they would not have missed them when they were not there. Whether we can imagine it or not, we should assume that, for the lifestyle to which they were adapted, "unconsciousness" was no great handicap. With these animals it was their behaviour itself, not their capacity to give an inner explanation of it, that mattered to their biological survival. As the occasion demanded they acted hungry, acted fearful, acted wishful and so on, and they were none the worse off for not having the feelings which might have told them why.

None the less, these animals were the ancestors of modern human beings. They were coming our way. Though their lives may once have been comparatively brutish and relatively short, as generations passed they began to live longer, their life histories grew more complicated, and their relationships with other members of their species became more dependent, more intimate, and at the same time more unsure. Sooner or later the capacity to explain themselves and to explain others – to take on, if it's not too grand a word, the role of a natural "psychologist", capable of understanding and predicting their own and others' behaviour within the social group – would become something they could no longer do without. At that stage would not their lack of consciousness have begun to tell against them?

Not necessarily. At least not at first, and not to the extent that all that's said above implies. For inner explanations are not the only kind of explanations of behaviour. Debarred as our unconscious ancestors may have been from looking in directly on the workings of their brains, they could still have observed behaviour from outside: they could have observed what went into the internal mechanism and what came out, and so have pieced together an external, objectively based explanatory model. "Why am I (Humphrey) looking in the larder?" Not, maybe, "Because I'm feeling hungry", but rather "Because it's five hours since Humphrey last had anything to eat" or "Because Humphrey has shown himself to be less fidgety after a snack".

In short, while our ancestors lacked the capacity to explain themselves by "introspection", there was nothing to stop them doing it by the methods of "behaviourism". "The behaviourist," wrote one of its first modern champions, J. B. Watson, "sweeps aside all medieval conceptions. He drops from his scientific vocabulary all subjective terms such as sensation, perception, image, desire, purpose, and even thinking and emotion." Who better placed to follow this recommendation than an unconscious creature for whom such conceptions could not have been farther from his mind? In fact, it is we conscious human beings who have trouble being hard-headed behaviourists: it is *we* who, as that other great behaviourist B. F. Skinner has lamented, "seem to have a kind of inside information about our behaviour. *We* have feelings about it. And what a diversion they have proved to be! . . . Feelings have proved to be one of the most fascinating attractions along the path of dalliance."

Why, then, when ignorance of the inner reasons for behaviour might have been bliss, did human beings ever become wise? Adam, the behavioural scientist, might with Newtonian detachment have simply sat back and watched the apple fall, but no, he ate it.

What tempted him was a leap in the complexity of social interaction, calling in its turn for a leap in the psychological understanding of oneself and others. Suddenly the old-time psychology that was good enough for our unconscious ancestors, which is still apparently good enough for Watson and Skinner, was no longer good enough for their descendants. Behaviourism could take a natural psychologist only so far. Human beings were destined to go farther.

At what point the threshold was crossed we cannot tell. But there is evidence that by three or four million years ago, and possibly much earlier, our ancestors had already embarked on what was in effect a new experiment in social living. Leaving behind the relatively dull life of their ape-like forebears – leaving behind their thick skins, large teeth and heavy bones, leaving behind their habitation in the forest and their hand-to-mouth existence as vegetarian gypsies – they sought this new life as hunter–gatherers on the African savanna. They sought it with stone tools, they sought it with fire; they pursued it with forks and hope. But above all, they sought it through the company of others of their kind. For it was membership of a cooperative social group that made the life of hunting and gathering on the plains a viable alternative to what had gone before. Life from now on was to be founded on collaboration, centred on a home base and a place in the community. This community of familiar souls would provide the context in which individuals could reap the rewards of cooperative enterprise, where they could benefit from mutual exchange of materials and ideas, and where (against all subsequent advice) they could become borrowers and lenders and then borrowers again – borrowers of time, of care, of goods and services. But more importantly, the community would provide them as they grew up first with a nursery and then with a kind of polytechnic school where they could learn from others the practical techniques on which the life of the hunter–gatherer depended.

But the intense social engagement which this new lifestyle entailed spelt trouble. For human beings would not, overnight, abandon self-interest in favour of the common good. While it is true that each individual stood to gain by preserving the social system as a whole, each continued also to have his own particular loyalties – to himself, to his kin and to his friends. A society based, as this was, on an unprecedented degree of interdependency, reciprocity and trust, was also a society which offered unprecedented opportunities for an individual to manoeuvre and outmanoeuvre others in the group.

Thus the scene was set for a long-running drama of personal and political intrigue. Men and women were headed to become actors in a human comedy, played out upon the flinty apron stage which formed their common home. It was a comedy which would be tragedy for some. It was a play of ambitions, jealousies, loves, hates, spites and charities, where success meant success in the conduct of personal relationships. When the curtain fell it was to those who, as natural psychologists, had shown the greatest insight into human nature that natural selection would give the biggest hand.

Imagine now two different kinds of player, with very different casts of mind. One the traditional unconscious behaviourist, who based his psychology entirely upon external observation, the other a new breed of introspectionist, who took the short cut of looking directly in upon the workings of his brain.

The behaviourist starts with a blank slate. In the manner familiar to those of us who have followed the progress of behaviourism as a modern science, he patiently collects evidence about what he sees happening to himself and other people, he correlates "stimuli" and "responses", he looks for "contingencies of reinforcement", he tries to infer the existence of "intervening variables", and thus, without prejudice, he searches for a pattern in it all.

This programme for doing psychology is not, let it be said, a hopeless one. It must have sufficed for our unconscious ancestors for many million years. It probably still suffices for most if not all non-human social animals alive today. With a bit of luck it might have sufficed for those who began to live the life of social human beings, had they but world enough and time, had there been no one else around with the gift of doing the job much better.

But now there was someone else around, and world, time and luck were all at once in short supply. An introspectionist had entered on the scene: someone who starts with a slate on which the explanatory pattern is already half sketched in. From earliest childhood the introspectionist has had the opportunity to observe the causal structure of his own behaviour emerging in full inner view: he has sensed the connection between stimulus and response, he has felt the positive and negative effects of reinforcement, he has been directly appraised of the intervening variables, and he has daily experienced the unifying presence of his conscious self.

In the first instance, certainly, the introspectionist's explanatory model applies only to his own behaviour not to others'. But once, in his own case, a pattern of connections has been forced on his attention, the idea of that pattern will dominate his perception in other cases where the connections are not openly on show. Once, in his own case, an outer effect has been seen to have an obvious inner cause, the idea of that cause will help him to make sense of situations where the effect alone can be observed. Notice that a fire in your own private hearth causes smoke to issue from your chimney, and try *not* to imagine that the smoke coming from the house across the road implies the presence of a fire within.

Thus the introspectionist's privileged picture of the inner reasons for his own behaviour is one which he will immediately and naturally project on other people. He can and will use his own experience to get inside other people's skins. And since the chances are that he himself is not in reality untypical of human beings in general – since the chances are that, just as from house to house there is generally no smoke without fire, so from person to person there is generally no looking in the larder without hunger, no running away without fear, no rage without anger etc, – this kind of imaginative projection gives him an explanatory scheme of remarkable generality and power.

Thomas Hobbes in *Leviathan* made the point, as few philosophers since then have dared to, "[Assuming on] the similitude of the thoughts and passions of one man to the thoughts and passions of another, whosoever looketh into himself, and considereth what he doth, when he does *think, opine, reason, hope, fear* &c and upon what grounds he shall thereby read and know what are the thoughts and passions of all other men upon the like occasions."

Let us return then to the age-old human play. Scattered among the population of unconscious behaviourists, there arose in time these conscious prodigies. Soon enough an unconscious Watson would find himself up against a conscious Iago, an unconscious Skinner would find himself laying suit to a conscious Portia. Natural selection was there to supervise their exits and their entrances.

It was clear where the story for the human species had to end. But for the rest of the animal kingdom? As the bias of my story must have shown, I am not convinced that any other species has followed the same path to consciousness as man. But studies of the social systems of other species are not far advanced, and studies of how individual animals themselves do their psychology have hardly begun. It may yet turn out that there are, in fact, non-human species whose social systems rival the complexity of man's; it may yet turn out that individuals of those species are, in fact, making use of explanatory systems which bear the hallmarks of a mind capable of looking in upon the inner workings of the brain. Stories have been wrong before. The cat, we know, does not walk by itself. But the rhino? Nothing suggests that the rhino gets inside another rhino's skin.

Meanwhile, for the obvious candidates – the social carnivores, the great apes – there will be biologists who in fairness want to leave the question undecided. Undecided, but not undecidable. In medieval England a jury could bring in four alternative verdicts at a trial: Guilty, Not Guilty, *Ignoramus* (we do not know), *Ignorabimus* (we *shall* not know).

"Ignorabimus" would be a counsel of philosophical despair. "Ignoramus" is the proper verdict for biologists. For if consciousness has evolved we shall know it by its works.

33

Human language – who needs it?

EMIL MENZEL

16 January 1975

Chimps can perform remarkable feats of communication with one another without the need to learn a language system based on human language. But such research has tended to be overshadowed by the popular appeal of attempts to teach chimps to communicate with humans.

Suppose you found yourself in an unknown land populated by strange creatures which are almost human in form, appear highly intelligent, live in well-organised social groups, but do not seem to share any verbal language or sign language in common with us. How would you go about studying their capacity for information processing and social communication? This problem would make for good science fiction but it is also the problem that faces any scientist who sets out to study chimpanzee communication.

There are at least three different solutions to this problem. Psychologists such as David Premack, Beatrice and R. Allen Gardner, and Roger Fouts would say, "Catch one of the beasts, preferably an infant, take it to the laboratory, and see if you can teach it English or some other system that meets the standards a human linguist would set for real communications." Naturalists would be more inclined to observe what self-trained animals already do with each other in ordinary circumstances. Some naturalists, such as Jane van Lawick-Goodall, Peter Marler and Jan van Hooff, say, "Watch every movement and listen to every sound that individuals make when they are in company with each other, and try to compile a dictionary of these natural communication signals."

The third group, myself included, would follow the lead of Karl von Frisch and Clarence Ray Carpenter, and say, "Look at the society as a whole and try to discover where it is headed and what are its goals. Then you can say what the individuals might have to communicate about, and ask how they do it." These three approaches to the general problem of communication are of course complementary rather than contradictory; here I shall concentrate on what has been learned by the last approach.

One of the most obvious goals of any mobile group of animals is to get to food without individuals getting lost from each other or being involved in perpetual conflict. After observing the spontaneous travel and foraging of chimpanzees for several years, I arrived at the following test procedure for analysing their accomplishments more closely. Six or eight young wild-born chimpanzees, who have lived together at least a year and who have come to form a very compatible and stable social unit, are locked together in a small cage on the border of a large outdoor enclosure. An experimenter then enters the enclosure and hides food or some other object in one or more randomly selected locations.

The next step is to take one of the chimpanzees out from the cage and carry him around the enclosure to show him the hiding places. After this the chimp is returned to his group and the experimenter leaves the scene. Several minutes later all of the chimpanzees are turned loose. Each time the test is repeated, a new set of hiding places is used. Each member of the group who can be carried away from his fellows without raising a big fuss is given the opportunity to serve as the informed animal or (as I call him here) the leader.

When he was released to look for the food the leader almost never searched a false location or missed a baited one by more than a metre or two or re-searched a place that he had already emptied of food. He also never used the same trail over which we had carried him, unless this was also a very efficient route. In other tests we found he could also remember the type and amount of food that he had seen in each place, and whether or not a snake or other frightening object was also hidden nearby. In sum, the chimpanzees seemed to know the nature and relative positions of most of the objects in the field, and their own position (at any given time) in this scaled frame of reference. As the late Edward Chace Tolman would have said, it was as if the leader had a "cognitive map" of the area and knew how to use it in planning an itinerary.

Figure 1 shows the results of one of the most complicated food-getting tests. I would guess that in this situation the chimpanzees' memory for places was better than my own. Also, obviously, they used a systematic pattern of foraging rather than running to and fro haphazardly and thus travelling a much greater distance than necessary.

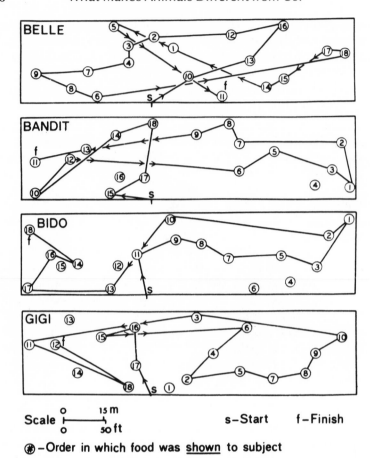

Scale

s–Start　　f–Finish

⊛ –Order in which food was <u>shown</u> to subject

Figure 1 *Maps showing the performance of each of the four leaders after they had been shown 18 different hidden piles of food. The connecting line shows the exact order in which the places were searched*

During the course of their travels, the chimpanzees usually moved as a very cohesive body and the informed animal clearly controlled the move. However, it was clear that the leader was by no means moving independently of his followers. If we tested him alone and thus gave him the opportunity to get all the loot for himself, he usually went nowhere, but begged at us with an extended hand or whimpered and tried to open the cage door to release his followers. If the chimps were tested together but for some reason the followers did not follow, the leader took a step or two and stopped and waited, glancing back from one animal to the next. He beckoned with a wave of the hand or a nod of the head, or tapped a preferred companion on the shoulder and "presented" his back to solicit "tandem walking" (with an arm around the waist). Sometimes he walked backward toward the goal while orienting toward the group and pursing his lips in a "pout face".

Polly and Bandit exchange glances as Bandit (extreme right) starts to lead the group towards a hidden goal, which he alone has seen, more than 50 m away

If all these devices failed, he whimpered and tried to pull preferred companions to their feet, he bit them on the neck until they started to move, or he dragged them along the ground by a leg. If they still did not follow, or tried to engage him in play, he rolled around on the ground screaming and tearing his hair. At this the followers usually ran to the leader, clung to him, and then started to groom him. Once they had placated him the leader usually "gave up" and no longer tried for the food unless someone else got up, started to walk away from the group, and in the correct direction. The leader then would rise too and travel behind the "followers" as long as their course was accurate. If they started to change course he grimaced and glanced back and forth at them and towards the goal location, or he ran to them, put a hand on their shoulder, and steered them back on course. Once the group got going well, first one animal and then another would glance at his fellows and step up the pace just a little. Soon the whole pack was racing down the field.

Usually the leader went in front, but this was by no means necessary. The followers knew which way to go anyway. Some followers in fact often ran several metres ahead of the leader, sighted back at him periodically, and then raced to search any likely looking hiding places which lay ahead on the leader's path. In tests where we left a pile of food only partially covered, so that it could be spotted from about 5 metres, and the leader was given less than complete knowledge of its location, one chimp even developed the strategy of running ahead of the leader and climbing a tree to scan the field from a height.

Did the followers know from the leader's behaviour that food, rather than something else, or perhaps just the desire for a stroll, was the object of the leader's travel? In some of our experiments we randomly showed the leader a toy or a snake or an empty pile. There was no question that not only the leader but also his followers knew what, if anything, was out there. Sometimes the leader led the group within 10 metres of a hidden snake and then hooted or barked an alarm, and would go no farther – at which a follower walked over toward the spot at which the leader stared, picked up a stick, and clubbed this place. If we had removed the snake after showing it to the leader (to avoid all possibility that the followers could perceive it for themselves), the animals, after mobbing the previous hiding place, climbed the trees or walked along the fence scanning outside the enclosure – exactly as they would search for a snake that they had all seen crawling along the ground only to disappear from sight.

In contrast, where the hidden object had been food (and we had removed it after showing it to the leader) the whole group searched the hiding place manually with no sign of caution, and with great excitement, as if they all expected food. If the leader had been shown an empty pile, he usually did nothing unless the others tried to get him going. In some instances the leader led the others over to the empty pile after they had tugged at him but when the group got over to the pile he sat down and watched as the others dived on the pile and searched through it.

In still other tests we found that if two leaders were shown different hidden goals the group would split up and the leader who was going to the better goal would get the larger following. For example, whichever leader was going to four pieces of food attracted about two followers to every one follower that accompanied

Belle (in the foreground) looks in the direction that her companion, Shadow, has 'pointed out'

the leader who was going to two pieces of food. The leader going to fruit (preferred) attracted more followers than the leader going to vegetables (non-preferred). The leader going to a new toy attracted more followers than the leader going to an old toy.

There was also a clear leadership hierarchy: if their goals were equal, some chimps consistently got a larger following than others. Generally speaking, the preferred leaders were more willing than the non-preferred leaders to share their spoils. Other characteristics which a chimpanzee had to possess before he attracted much of a following were familiarity (strangers were shunned) and ability to move out independently from others (with perhaps an occasional glance backwards to make sure one is being followed).

Could the leader tell the others where to go without actually going there himself? On some occasions a leader started out only to give up for one reason or another and the followers took over and continued 50 metres or more in the correct direction. However, when we tried to test this behaviour formally, by putting the leader with the group for a few minutes and then taking him out and putting him in a separate cage while the followers were turned loose, the results were (as we had predicted) inconclusive. The leader whimpered and the followers went nowhere, but tried instead to open the cage door and release him. I am more inclined to attribute this result to motivational factors than to any failure of communication.

I base this statement on experiments in which we did not show the leader the hidden object, but only gave him a social cue as to its direction. For example, we merely pointed in the appropriate direction manually; or, to use signals that were more like the ones that the chimpanzees themselves used, we took a few steps and leaned forward and acted as if we had seen food. In some experiments we used two piles of food of different size, and used a long walk as a cue for the big food pile and a short walk as a cue for the small pile.

Even though the experimenter had left the enclosure and there had been several minutes delay between his signal and the opportunity to respond, the leader usually set out within 10 degrees of the indicated course. Walking, pointing, and merely orienting visually in a fixed direction were almost equally effective cues. On tests where two food piles were involved the leaders almost always went to the larger pile first and, as in most of our tests, their performance was excellent even on the first trial. Control tests, in which we put no food at all or otherwise gave the cue in a false direction, ruled out the possibility that the animals could smell the food or were using cues other than the ones we had intended.

Exactly how far ahead of us in space the leaders could extrapolate I do not know, because our enclosure was too small to test the upper limits of their ability but it is no less than 80 metres. As for the time limit of their ability to remember a social cue of direction, I would estimate it is at least a half hour.

It is, of course, easier to convey directional information than it is to convey the exact location (direction and distance) of an object. However, the chimpanzees did appear to use several different types of cues to pinpoint the exact location. To any student of spatial perception, the most interesting cue was triangulation. If the same animal successively oriented toward the same point in space from several different places (or, alternatively, if several group members simultaneously oriented toward the same point from several different angles) some followers seemed to put these several directional signals together and immediately infer the exact location of the goal.

In sum, the chimpanzees were able, and without any deliberate training on our part, to convey to each other the presence, direction, probable location, and the relative desirability and undesirability (if not the more precise nature) of a distant, hidden goal which no one had directly seen for himself. From my discussion with other primate research workers, particularly Hans Kummer, van Lawick-Goodall, Robert Hinde, and their students, I would be very surprised if at least this much could not be demonstrated for many wild monkeys and apes, if not for pack-living carnivores and for many other animals of different types.

Why have primate field studies thus far reported little or no evidence for communication about the environment? Largely, I suspect, because these studies have not identified or controlled the goals the animals are responding to, and because, under the influence of Lorenzian ethology, they have restricted themselves to describing the particular signals that individuals make and trying to compile a sort of dictionary of these signals. A dictionary of signals (or "elements") does not necessarily tell us what messages are actually conveyed across a group, for the same signals can take on a totally different meaning according to who gives it and in what context it is given. The same message can be conveyed with many different ritualised signals or even with none at all, but merely a glance or a bodily "point" or a nod of the head in the appropriate set of circumstances.

Perhaps various species differ less in how much information individuals can transmit to each other than in how and under what conditions they do it. A big difference between the signal systems of chimpanzees and the signal systems of bees and ants is that chimpanzee leaders do not restrict themselves to any small set of inborn or highly ritualised signals. They can learn to do almost anything they have to do to achieve their overall objective: to get to distant, hidden goals without having to go it alone. Similarly, followers do not have to attend to any one particular signal or nuance of the leader; they need only attend to the general situation closely enough to get the general message. Biologically speaking, it does not matter whether the follower gets his cues from visual, auditory or some other sensory channel, or from the leader, another follower, the environment, or some aspect of interaction thereof. The important thing is that he, too, is able to get to the distant resources without getting lost from the group. And, except in experimentally contrived test situations, there are always many alternative, redundant sources of information regarding the group's goals. Gestural or dance languages could no doubt be learned by wild chimpanzees, but they would still be necessary only to the extent that alternative cues or information processing systems were not sufficient to remove the mismatch between what the leader knows (or desires) and what the follower knows (or desires).

I know of no evidence to date that chimpanzees trained in a humanoid language but otherwise deprived of experience with other chimpanzees – as, for example, in the studies of Premack and

Belle cautiously uncovers a hidden model of a snake, which had previously been shown only to Bandit (second from right). If the hidden object had been food, it would have been uncovered by hand

Beatrice and R. Allen Gardner — could tell each other any more than wild chimpanzees are capable of conveying to each other with their "language of the eyes". In my group of chimpanzees, the most dramatic and humanoid-looking signals were made by the most infantile and least efficient leaders, and they decreased markedly as the animals gained experience at leading other chimpanzees.

The leaders who best conveyed the nature, direction and relative desirability of their goal were those who simply took a few steps "independently" and then glanced back to see what the others were doing; and, finally, they abandoned their own goal and joined another potential leader if he clearly seemed to be more eager and well oriented than they. As Norbert Wiener said, the ability that two animals have to pick out the moments of each other's special active attention, and to use these moments as cues to the nature of the environment, is itself a "language" as varied in possibilities as the range of impressions that the two animals are able to encompass.

Does the chimpanzee's "natural" system for communicating about the environment reduce, then, merely to unintentional displays of eagerness on the part of the leader, together with the innate tendency of others to follow whoever goes the fastest? I think not. I found in my studies, and van Lawick-Goodall has found in wild chimpanzees, that older animals can actively inhibit most signs of emotion if they so choose, and thus withhold information from each other, if not actually lie. For example, a chimpanzee who has seen food never "automatically" runs to it, grabs for it and stuffs it in his mouth. If a stranger or a dominant is nearby, he waits as long as he has to until the coast is clear, or he might even get up, lead the other animal somewhere else, and then (while the follower is otherwise engaged) circle back for the hidden food. Conversely, followers are sometimes very acute in assessing from a deceptive leader's displacement activities and holdings-back what he is up to. The less eager and obvious he *tries* to act, the more closely the followers might keep him under surveillance.

Particularly in the light of what other researchers have disclosed in the last five years about chimpanzee and human information processing, the observations I have described here suggest that chimpanzees perceive the world and interpret each other's behaviour in ways that are not ridiculously different from the ways that we ourselves use, especially when we are silent and non-sedentary. Whether or not untrained chimpanzees have real language as a linguist would define it, they do possess information processing systems, predominantly visually based ones, which are to a considerable degree one of the same form as our own verbal language and which serve the same biological functions.

As Premack has said, it is as if all of the cognitive structures necessary for grammar are already there not only in preverbal children but also in nonverbal chimpanzees. These subjects do not have to be taught grammatical modes of thought, in the ordinary sense of the word "taught", but only provided with a means for expressing their knowledge to us, their observers, in terms that we can understand. By human standards, or even by the standards of language-trained chimpanzees, self-trained chimpanzees do not *seem* to have a great deal to say, but how much of this result is attributable to their poor communicative ability and how much is due to our limitations as observers is still an open question.

34

Washoe the ape learns to talk

'MONITOR'

28 August 1969

The first successful attempt to teach a chimp to use a language system
employed a language that should come easily to the expressive ape –
the American Sign Language.

"He understands every word I say" is a common boast among dog
owners, particularly elderly female providers for pekinese. Taken
with the required large pinch of salt, there remains something in
what they say: dogs, and other intelligent animals, are able to
acquire a fairly extensive vocabulary of words they can recognise.
They cannot, however, answer back. All attempts to teach animals
to use human language for communication have failed dismally,
including heroic efforts made in the United States by the husband
and wife team of the Hayeses, who tried to teach their chimpanzee
Vicki to talk by bringing it up as if it were a human baby. In six
years, Vicki learned only four sounds that approximated to words.

But for the last three years, another dramatic experiment with a
young chimpanzee has been taking place in the USA, at the University of Nevada and there a quite remarkable breakthrough has
been made. Up to a period ending in December last year, Washoe –
the young female chimp in question – had acquired a vocabulary of
some 60 words, and was able to put together combinations of these
words into simple sentences. But Washoe does not speak in the
conventional sense; she uses her hands in a sign language of great,
and accelerating, complexity. For the first time, human beings have
established limited two-way communication with a member of
another species.

The human beings in question are R. Allen Gardner and Beatrice
T. Gardner, husband and wife, and professor and research associate
respectively in the psychology department of the University of

Nevada. They chose a chimp for their experiment in communication because of that species known high intelligence and sociability. The decision not to attempt to teach it to talk was based on the failure of previous attempts, and the recent anatomical studies which have shown that chimpanzees are physiologically unable to form the wide and subtle range of sounds of human speech. The choice of the sign language approach was made for the simple reason that the Gardners wanted to exploit for communication something that chimps are naturally good at – and gesturing is such a response. The particular sign language chosen is that employed by the deaf in North America, known as American Sign Language (ASL).

Washoe herself was caught in the wild, and was between 8 and 14 months old when obtained by the Gardners. At this age, chimps are normally wholly dependent upon their parents; they become semidependent between the age of two and four years, mature sexually at eight and have a life-span of well over 40 years. The experiment, which began in June 1966, was conducted largely in a special home constructed for Washoe in the psychology laboratory, with occasional outings in the car and visits to the Gardners' home. The chimp's confinement was minimal; the intention was that all the humans she encountered should be her friends and playmates as well as providers and protectors. As far as possible, these humans conversed among each other, when in Washoe's presence, in ASL; the intention was to avoid giving the impression that "big chimps talk and only little chimps sign".

The Gardners described the first 22 months of the experiment – up to March 1968 – in detail in last week's *Science* (vol. 165, p. 664). Washoe learned by three principal methods: imitation, spontaneous manual "babbling", and as a result of instrumental conditioning. Imitation has mainly been useful for improving Washoe's "diction", and for increasing the frequency of her signs, rather than introducing new ones. If the chimp failed to use an appropriate sign, in a given situation, or made a rather poor version of the sign, her human companion made an extravagantly correct version of the sign for Washoe to imitate – which she would unless pressed too hard (when she might throw a tantrum).

The young chimpanzee was continually exposed to a wide variety of activities and objects, and at the same time shown the sign for each, in the hope that she would come to associate sign with object (or activity) and come to make the sign herself. This she did, often after a very long delay. For instance, the sign for "toothbrush" was given to her after every meal while she was put through the (to her) distasteful ritual of teeth-cleaning. On a visit to the Gardner home, in the tenth month of the experiment, Washoe climbed on to the washbasin and made the sign for toothbursh (an index finger rubbing the teeth) when she saw the family's collection. This sign was quite spontaneous, and was apparently made for no other motive than simply the desire to communicate.

As human babies babble, and are encouraged to do so by "goo-goo" noises from their parents, so Washoe "babbled" with her arms and hands, and was encouraged in this by her human companions' clapping, smiling, and repeating the gesture. Her babbling was increased during the experiment, and more and more she has tended to indulge in a wild flurry of random arm-wavings when "stuck for a word". If the babbled gesture has approximated an ASL sign, then her mentor makes the correct version of the sign and attempts to indulge in the activity it describes. Washoe probable learned the sign for "funny" in this way. The Gardners have also allowed Washoe to incorporate into her vocabulary some signs of her own, such as "come-gimme" (a beckoning gesture) and "hurry" (shaking her open hand vigorously from the wrist).

Finally, some of Washoe's vocabulary was taught her by straightforward conditioning techniques, in which she was rewarded by tickling (a thing for which all young chimps have a passion) when she made an appropriate gesture. This gesture was commonly shown her by holding her hands, and guiding them through the required motion.

Several points about the experiment and its results are extremely promising. First, there is the rate at which Washoe acquired new signs – a rate which was increased almost exponentially. In the initial seven months of training, she learned four signs; in the second seven months nine new signs appeared; 21 new signs turned up in the next seven months; and, according to a report in the *Washington Post*, more than another 30 signs were acquired in the succeeding period up to December of last year. Next, her ability to differentiate has increased. The sign for "flower", for instance, frequently appeared when Washoe smelt something – cooking perhaps. Now, she has separate signs for flower and smell, and only rarely confuses them. More important, Washoe is gaining the extremely sophisticated ability of being able to transfer a sign from the specific to the general; thus a sign learned for a flower is now used by Washoe to cover all flowers, indoors or outdoors, real or in pictures.

Lastly, and most significant of all, the young chimp is now able to combine signs into simple phrases – for instance, "open food drink" for open the refrigerator, and "listen eat" at the sound of an alarm clock signalling mealtimes. In the period after that reported in the *Science* article, Washoe has also learned the pronouns "I-me" and "you", which she is employing in short sentences.

The chimp's intellectual immaturity (she was still, at the end of the first 22 months of experiment, little more than an "infant"), the fact that she can transfer and combine signs, and the acceleration in her progress, all point to her still having a long way to go. The Gardners' hope is that Washoe can be brought to a point where she can describe events and observations, in an extended series of signs, to an observer who has no other source of information. The psychological – indeed, philosophical – implications of such an achievement would be quite extraordinary.

35

How do you recognise a
talking chimp?

'MONITOR'

14 October 1976

Three American laboratories could report that their chimps were skilled communicators using three different language-based systems. But was it language?

The search for the origins of spoken language is leading biologists to teach sign language to chimps at one extreme and to sophisticated computer analysis of speech at the other. Progress in this fascinating enterprise was one of the highlights of the Sixth Congress of the International Primatological Society held recently in Cambridge, England.

The mentors of two of the three most famous talking chimps were present; David Premack, who taught Sarah to use plastic "words", and Duane Rumbaugh, who taught Lana to use a computer-based system of symbols. The Gardners, who taught American Sign Language to Washoe could not attend, due to the untimely death of one of their protégées. The session also benefited from the presence of Peter Seuren, a dyed-in-the-wool linguist, Emil Menzel, an expert on natural communication in chimps (see p. 244), and Elena Lieven, a psychologist working on language development in children. The stage was set for an appraisal of the evidence and a re-examination of goals.

Lana now gets all she needs, including human companionship and conversation, by talking to a computer. Her words are coloured diagrams on the surface of keys, and the computer detects and responds to, her grammatical use of these lexigrams. Rumbaugh's attitude is that chimps can master aspects of human language. For example they can change word order in a sentence while preserving meaning. They also coin new words appropriately. For instance, Washoe calls a brazil nut a "rock berry", and refers to ducks as "water birds". Chimps are clearly capable of learning something. Whether this is admitted as language is largely a matter of definition. If the linguists would just stick to a particular definition, Rumbaugh suggests, he would show that Lana had language.

David Premack attempted to supply a definition of language. His own work consists of isolating features of language and getting chimps to master these, but he admitted that he would never know when the list of features was complete. Three possible subdivisions of language are syntax, communication, and representation. Chimps do string symbols together according to rules, but there is no evidence that the rules are anything like the rules of human syntax. The evidence for communication is better, and even provides glimpses of the chimp's intentions. It is when we get to representation, the use of symbols for properties of the real world, that we find chimps are very good indeed. Premack seemed to be saying that, with the right training, chimps definitely have the capacity to learn something. If it is not language, what is it?

The issue of whether what chimps can do is or isn't language was side-stepped by Emil Menzel. He was more concerned with the reasons why they are able to do what they can. The capability of the species far exceeds its known behaviour in the wild, and this raises again the old issue of evolutionary pre-adaptation. The problem here is that we don't really know whether chimps use all their capabilities in the wild, and we may never know. After all, as Menzel pointed out, you can watch people for a long time and never see someone inventing the calculus. Then again, if chimps are not in fact using their full capabilities, one would want to know why.

In general the chimps' representatives in the chimp language debate agreed on two important points. Given intensive training chimps, and other apes, certainly have the capacity to learn to use systems remarkably like language. But, and this is perplexing, they do not appear to do so in their natural state. Have they no need?

It was Peter Seuren who showed us why the "is it or isn't it" debate is futile. Linguists, he said, are simply not equipped to talk about non-human language. All the apparatus of linguistics has been developed to investigate and describe the features of various human languages. Linguists, he admitted, lack the means to place systems of communication on an ordered continuum of "language-ness". The direction Seuren favours is to accept that apes can learn a type of language, and to explore the limits of their linguistic ability.

For example, all human languages possess certain universals. Chimps have already shown they can master the simpler ones, and

Seuren would like to extend this research to encompass ever more complex universals. In this way we might be able to develop a feel for language as a continuum, and be better able to study it as a result.

Elena Lieven and John McShane put the debate in the context of human language development. In particular they stressed the need to view language within the social context of the animal. Thus language is seen not as an end in itself, but as a means of achieving certain goals. The young child's utterances can be understood only in the context in which they occur, and may serve to support conversation and communication long before they have full meaning.

Primates, because they learn language slowly, and because the language they use is simple, are of great use in linguistics. They provide insights into the normal pattern of human language development. For example, it is now relevant to ask whether symbolic play, the use of objects to represent other objects in play, is a necessary precursor of language in primates. Piaget claimed that symbolic play was essential for humans, but conclusive evidence either way has never been forthcoming, because all children show symbolic play and all children learn language. Primates could sort this point out. More importantly, the programmes devised for teaching language to chimps have been used, with great success, to provide severely handicapped people with a means of communication.

36

And now . . . the thinking pigeon

HERBERT TERRACE
4 March 1982*

Pigeons and rats are not normally regarded as the Einsteins of the animal kingdom. But recent experiments show that even these humble laboratory animals succeed in tasks that apparently require considerable brainpower. And they do it without language.

Human thinking about animal minds is sharply divided. For many people it is self-evident that animals have highly developed mental abilities. How else can one explain an animal's ability to navigate huge distances, to solve difficult problems posed by research psychologists or to communicate about mating, territory, the presence of predators, the location of food, and so on? Indeed, some psychologists have interpreted these and other instances of complex animal behaviour as evidence that animals share with human beings a consciousness of their feelings and of significant events in their external world.

A moment's thought will show that a belief in the existence of animal minds follows directly from the logic of the theory of evolution. In *The Expression of Emotion in Man and Animals,* Darwin noted explicitly that the assumption that mental powers evolved gradually implies the existence of precursors of the human mind in man's near and not-so-near relatives. It is important to recall, however, that Darwin's view of continuity between the mental abilities of animals and human beings created much controversy. At issue was the sharp departure of Darwin's theories from a tradition that emphasised man's uniqueness as a thoughtful and rational creature. Supporters of the latter position have regarded animals as unconscious brutes whose behaviour, however complex, is pushed and pulled automatically by cues emanating from their internal or external environments.

* This article is based on an article by the same author that appeared in *New Society* in 1982.

In his well-known analysis of the differences between human and animal behaviour, René Descartes argued that animal behaviour was the direct product of biological machinery. He felt confident that the most complicated types of animal behaviour could be reduced, without residue, to some configuration of reflexes in which thought played no role. As we shall see shortly, Descartes' position has, until recently, been vindicated by modern studies of animal behaviour.

According to Descartes (and to many philosophers and psychologists who have adopted his point of view), the ability to think presupposes an ability to use language. While animals engage in reflexive acts of communication, the form and meaning of such acts are rigidly fixed. One bird cannot learn the song of another species nor can it sing a mating song outside the well-defined circumstances of mating. In contrast, language allows humans to communicate in arbitrary ways that are beyond the reach of the most intelligent animals. The arbitrary nature of human languages is most readily revealed by their huge variety. "Eat", "manger", "essen", "mangiare", and so on, all convey the same meaning. More impressive is the human capacity to combine words into phrases and sentences in order to create new meanings that individual words, in isolation from one another, cannot convey.

Recent projects that tried to teach apes some of the fundamental features of human language have had little success. Early evidence that apes had created sentences dissolved in the face of discoveries that the apes' language-like behaviour could be explained as rote learning and/or imitation. Those talents, however strongly developed, do not provide the competence needed to master a language.

The failure to demonstrate linguistic competence in animals does not imply that animals cannot think. There remains the logical possibility that thinking can occur without language. At least in the case of human thought, the role of language is not readily discernible in the case of visual or musical imagination or during the recognition of pictures or melodies. Language may nevertheless play an indirect role in such thoughts in that thinking without language may presuppose a mastery of the symbolic skills needed to learn a language. In the absence of examples of human beings who have not learned some language, but who can nevertheless think, it is difficult to make an air-tight case that thinking does not presuppose some ability to use language.

As I noted earlier, one cannot appeal to language in characterising the thought processes of an animal. Thus, animals may provide an invaluable opportunity for studying thought without the complication of language. Before that opportunity can be realised, however, it is necessary to find clear-cut instances of animal behaviour that do not allow the kinds of mechanical explanations suggested by Descartes. That task has been made all the more difficult by modern extensions of I. P. Pavlov's original reflex model of behaviour – extensions that provide reflex models of *learned* (as opposed to inborn) behaviour (see p. 22).

One of the champions of reflex models of behaviour, B. F. Skinner, developed techniques for "shaping" the desired behaviour with a minimum of error. First he rewarded (reinforced) approximations of the response he sought to condition. From those approximations he was able to select and strengthen the desired response.

As an example of the shaping process, consider Skinner's technique for conditioning a pigeon to peck a small, illuminated disc. The pigeon is placed in a Skinner box, an experimental chamber that insulates the pigeon from visual and auditory distractions. The Skinner box works on the same principle as a vending machine. After the animal makes the appropriate response, the desired reward appears. In the case of the pigeon, a peck directed at the lighted disc provides a few seconds' access to a food hopper.

When the pigeon is first placed in the chamber it is taught to eat from the food hopper. Since it shows no tendency to peck the response disc, the trainer rewards the pigeon for orienting in the direction of the response disc. Next, he requires the pigeon to approach the disc and to position itself directly in front of it. Once the pigeon complies, the trainer waits for the pigeon to make a small pecking movement towards the disc. By gradually raising the response requirement, from weak to strong pecks, the trainer is able to

Perhaps pigeons are not so featherbrained after all . . .

"shape" the pigeon to peck the disc. Since that response automatically operates the feeding device, the pigeon will continue to peck until it loses interest in food.

The versatility of Skinner's conditioning procedure was enhanced greatly by a variety of important modifications. By restricting the availability of food to those intervals when a particular stimulus is present, Skinner was able to train fine discriminations. Suppose, for example, that the colour of the response disc alternated between green and blue and that pecking was rewarded only when the disc was green. Under those circumstances the pigeon would peck mainly when the disc was green. Another requirement for reward might be the passage of a particular interval of time. Pecking, for instance, might be rewarded only after a minute has elapsed since the previous reward. Under those circumstances, the pigeon will not peck very much immediately after it is given a reward. As the minute runs its end, however, its rate of pecking increases rapidly.

Through the use of discrimative stimuli, "schedules" of reinforcement and other elaborations of his basic conditioning procedures, Skinner was able to train complex forms of animal behaviour. His successful efforts with animals prompted him to conclude that the same principles would enable him to teach a child to read and write, to play tennis and to act politely at a party. Skinner would argue that the behaviour of a pigeon who is rewarded only when the response disc is green and only after a minute has elapsed since the last reward is similar to the behaviour of a human commuter waiting for a bus. First, the commuter must locate (discriminate) the appropriate stop for a "green line" bus. Having done that, our imaginary traveller looks more and more frequently in the direction in which the bus is expected to appear as the bus's scheduled time of arrival approaches.

While Skinner's conditioning paradigm was different from Pavlov's, the behaviour it established was just as thoughtless and mechanical. In combination, the two types of conditioned response (called "classical" and "operant") provided a formidable array of explanatory devices that could be applied to virtually any kind of animal or human behaviour. However thoughtful some behaviour may seem, a behaviourist would find some way of reducing it to an instance of conditioned behaviour.

During the past 20 years, the view that principles of conditioning can explain human behaviour has been found wanting for a number of reasons. Noam Chomsky, and other linguists, have revealed in great detail the inadequacies of behaviouristic explanations of language. Similar problems, which have been noted in the case of human memory, provided an important impetus for the growth of cognitive theories of human behaviour. Such theories of complex human behaviour did little, however, to help the cause of those psychologists who wanted to establish that animals can think. In the case of animals it remained to be shown that a purported example of thoughtful behaviour was not a thoughtless reaction to some stimulus, however subtle.

Approximately 70 years ago, Walter Hunter defined quite clearly the circumstances that would suffice to show that an animal could represent (that is, think about) some event in its environment. His strategy was to show that an animal could perform some task *without* the benefit of an eliciting or discriminative stimulus. Hunter observed, "If comparative psychology is to postulate a representative fact . . . it is necessary that the stimulus represented be absent at the moment of the response. If it is not absent, the reaction may be stated in sensory-motor terms."

Defining the conditions under which one could appeal to the representational powers of an animal turned out to be much easier than collecting evidence which satisfies those conditions. What proved elusive until recently were unambiguous examples of behaviour that, *in principle*, could not be controlled by some environmental stimulus. Even when the identity of the controlling stimulus could not be specified, a behaviourist might hypothesise the existence of such a stimulus. In fairness to the behaviourist's position, I should note that such hypothetical stimuli often proved to be all too real.

The story of the German horse, Clever Hans, provides a classic example of how the discovery of previously unsuspected cues led to the downgrading of an animal's purported ability to solve arithmetic and other equally demanding problems. The problems were written as questions (eg, $2 \times 3 = ?$) on a blackboard by Hans's trainer (an honest man who believed genuinely in his horse's mental powers). Hans answered these questions by tapping his right front leg an appropriate number of times. Careful observation later revealed that Hans's trainer unwittingly cued the horse as to when to start and when to stop tapping his leg. After posing the problem, the trainer inhaled deeply, uncertain as to whether Hans would respond appropriately. That was the start cue. Once Hans tapped enough times, the trainer exhaled out of nervous relief. That was the stop cue.

Other less dramatic examples of uncontrolled cues reveal how difficult it is to defeat hypotheses about their potential contribution to the solution of some problem. Rats running a maze may pick up faint odour trails left by rats who had previously run the maze. In a discrimination task, the failure to change the position of the stimuli from trial to trial could lead to erroneous conclusions that an animal has learned a complex discrimination when, in fact, it learned the simpler task of responding to a particular location.

Recent research on animal cognition has, for the first time, shown that an animal can solve certain kinds of problems without relying on environmental stimuli, hypothetical or real. In order to do so, the animal must create its own internal representations of past events and use those representations in formulating a solution to the problem on hand. In several experiments now, Hunter's strategy of requiring an animal to use representations of its previous experience has revealed the existence of sophisticated forms of thinking in a variety of animals.

A recent experiment I have performed shows that pigeons can learn arbitrary sequences of responses in much the same way that a child learns to memorise a nursery rhyme or a telephone number. While these sequences learned by rote have no meaning, they must be produced from memory without the help of step-by-step guidance from environmental cues.

In an experimental chamber of the kind I described earlier, I confronted a pigeon with four response discs each a different colour. In order to obtain food, the pigeon was required to peck the colours in a particular sequence regardless of the configuration of the colours. Suppose that the sequence the pigeon was required to produce was: red(A)–green(B)–blue(C)–yellow(D). On one trial the configuration of colours on the response discs might be C, B, A, D. On the next trial it might be A, C, B, D, and so on. Whatever the configuration of the colours, I rewarded the pigeon if, and only if, it pecked the colours in the sequence A–B–C–D.

After learning to peck the required sequence on a fixed set of configurations, I tested the pigeons on novel configurations. On the first trial they responded just as accurately to the novel configurations as they did to the training configurations. This shows that they did not perform the required task by memorising a list of particular configurations of the four colours.

Even more impressive was the pigeons' performance on simultaneously presented pairs of colours drawn from the original "list" of four colours. On some trials the pigeon might be shown the colours A and B, on others the colours B and D, on others A and D, and so on. That the pigeons performed accurately to arrays consisting of A and B is hardly surprising. When shown the colours A, B, C and D, the pigeon was required to peck A first and B second. What was surprising was the pigeon's ability to respond just as accurately to arrays consisting of the colours B and D as it did to arrays consisting of the colours A and B. Arrays consisting of B and D provided the pigeon with neither the benefit of the normal starting colour (A) nor the benefit of an adjacent element (C) after it responds to B. The pigeon's performance on this task is comparable to a child being able to recite in the correct order, the second and fourth lines of a nursery rhyme or to play non-adjacent notes of a musical scale. Such performance is possible only if the child has a representation of the sequence from which the two lines or the two notes are drawn.

Just as children can produce particular melodies, they can also discriminate one melody from another. Pigeons are just as skilful. The task is to look at a sequence of three colours (A, B, and C) and to respond to one response disc (a "yes" response) if the colours appear in the sequence A–B–C and to a second response disc (a "no" response) if they appear in any other sequence. Pigeons can discriminate such sequences fairly readily, even though they have to make their "yes" and "no" choices after the last colour of the sequence has been presented.

A pigeon's ability to discriminate sequences demonstrates two important mental abilities. One is analysing a sequence of arbitrary elements, presented one at a time. Notice, for example, that the pigeon cannot make a correct choice simply by attending to the last element of the sequence. If it used that strategy, it could not differentiate the sequences A–B–C and B–A–C. The pigeon must also "keep in mind", in the absence of any external stimuli, the sequence it saw until the "yes" and "no" choices were presented.

Another study of animal memory (performed by David Olton) shows how rats can reflect upon recent experience so as to find different sources of food with a minimum expenditure of effort. Consider the problem faced by a hungry rat who is placed in the centre of a "radial" maze – a maze that consists of a central circular platform and (typically) eight identical arms radiating from that platform. Each arm, which is separated from its neighbouring arms by 45 degrees, contains a single pellet of food at the start of the trial. Once the rat feels comfortable in its new surroundings, it quickly finds and eats each of the pellets. What is unusual about the rat's behaviour is the efficiency with which it obtains the pellets at the end of each arm. After a few days it learns how to empty the maze with but a *single* visit to each arm and without relying on entering the eight arms in a fixed sequence. On one trial the rat's sequence might be 7–8–4–1–5–3–6–2, on another it might be 4–2–3–7–5–1–8–6, and so on. In short, it is impossible to predict the rat's sequence on one trial from the sequence that it followed on previous trials. Such performance poses an interesting problem: how does the rat remember (in some experiments for several hours) which arms it had visited previously?

One possibility is that, as the rat enters or leaves an arm, it leaves some kind of cue that it can detect when it returns to the entrance of that arm. A clever variation of the basic training procedure showed that this was not the case. After making four choices, Olton confined the rat on the centre platform. He then rotated the arms of the maze some arbitrary distance, say 90 degrees. Upon its release from the centre platform the rat had the following choice: either to avoid arms it had visited previously, no matter where they were located, or to visit new spatial locations, even if that entailed entering an arm it had previously visited. Most rats chose previously unvisited *spatial* locations even when this meant that they had to re-enter one or more arms they had already visited.

Whatever system the rat uses to remember which arms of the radial maze it has visited, it should be clear that its choices cannot be attributed to an external cue. It is true that the rat must make use of features of the room containing the maze; for example, the position of an overhead light, the direction of exhausted air, odour gradients, and so on. However, it should also be evident that the rat's ability to orient itself with respect to these landmarks cannot, by itself, explain the efficiency of its performance in the radial maze. The rat must not only orient itself in the maze but it must also *represent* the locations of those alleys it has already entered. Whatever the form of that self-generated representation, it cannot be regarded as a conditioned response to some feature of the environment.

Animals cannot only remember places they have visited but they can also remember what they have seen. A memory test, devised by A. Wright, exploits a monkey's ability to judge whether two photographs are the same or different. At the outset the trainer shows the monkey a pair of photographs for a brief period of time (about one second). On a randomly selected half of the trials, both members of the pair of photographs are the same; on the remaining half they are different. When both photographs are the same the trainer rewards the monkey for pressing a "same" button. When the photographs are different the monkey is required to press a "different" button. A monkey can be trained to respond correctly on this task even when faced with novel pairs of photos on each trial.

During the second phase of training the monkey sees the members of each pair of photographs *successively*. Despite intervals of 10–20 seconds between the appearance of the first and the second members of each pair of photographs, the monkey responds highly accurately. During the final phase of training, the monkey looks at *lists* of as many as 20 photographs before the test photograph is presented to him. There are different lists of photos on each trial. As previously, the task is to respond "same" if the test photograph appeared in the list and "different" if it did not. Even under these conditions the monkey's performance is highly accurate.

In order to respond correctly during the second and third phases of training, the monkey must make mental comparisons between representations of photographs it has seen earlier during the trial and the test photograph. Accordingly, Wright's monkeys provide solid evidence that monkeys can think about stimuli that are not physically present. A similar, but less obvious, argument is needed to explain the monkey's performance during the first phase of training, when the pairs of photographs were presented simultaneously. The novelty of each pair of photographs precludes the possibility that the monkeys learned to respond to particular features of each photograph. In order to make judgements of sameness or difference, the monkeys had to rely on their own abstract concept of same and different, a concept that, by definition, transcends learning to discriminate features of particular photos.

While psychologists have yet to devise a rigorous definition of human intelligence, there is little doubt that the ability to form concepts is a basic factor. As children, we have to learn that cows and sheep are animals, that playthings are inanimate, that the same colour name can apply to objects that are otherwise different, and so on. As we get older our concepts become more elaborate. Just the same, the process of learning concepts remains the same: finding likenesses in the varieties of seemingly dissimilar objects and creatures.

Children seem to acquire concepts easily; how they do so is another question. One problem is the difficulty of capturing all exemplars of a particular concept by a simple rule. Consider, for example, such seemingly simple concepts as "chair" and "tree". While certain exemplars of these concepts are obvious, borderline cases such as stools or certain kinds of shrubbery are less so. To clarify this problem, ask yourself how one could instruct a computer to discriminate photographs of trees (elm trees, willow trees, palm trees, etc.) from photographs of non-trees (treeless mountains, animals or sunsets, for example). Using optical scanning devices (such as those used to "read" postal codes on envelopes), the computer can assimilate any feature of these photographs. Can the computer be programmed to discriminate trees from non-trees?

It turns out that the most powerful computers now available cannot discriminate all instances of trees, all instances of chairs, and so on. While it is not difficult to write rules that will accommodate most of the obvious examples of such concepts, the computer makes errors when confronted with borderline cases that pose no problems for humans (e.g. celery, beach grass, rocking chairs, and so on).

Our inability to write rigorous rules for most of the concepts we form makes all the more impressive the results of studies performed by R. J. Herrnstein which show that pigeons can readily master a concept such as "tree". The procedure for teaching pigeons to form a concept is quite simple and at least partially familiar. During the first stage of training, any peck at the response disc will activate the feeder. Once the pigeon pecks the disc reliably, colour photographs are back-projected onto a small screen above the response disc. If the photograph contains one or more trees, some fraction of the pigeon's pecks activates the feeding device. The pigeon never obtains a reward for pecking if the scene contains no trees. A new scene appears every 30 seconds. Since the slides containing trees alternate irregularly with those that do not, the only basis the pigeon has for predicting when its pecks will operate the feeder is the presence of a tree in the picture above the response disc.

Within a few days, the pigeon reliably pecks more rapidly when trees are projected than when they are not. In order to show that the pigeon's knowledge about trees is not limited to the slides with which it was trained, the tester occasionally presents novel slides, some with trees and some without. By responding to the test slides as accurately as they responded to the training slides, the pigeons demonstrated that they had truly mastered the concept of "tree".

Similar experiments were successful in training concepts such as "water", "fish", "pigeons", "people", and, indeed, a particular person. The concept of water was exemplified in pictures of natural bodies of water, pools, snow, ice, raindrops, and so on. Learning the concept of a particular person entailed recognising that person in close-ups, in long shots, in summer and winter dress, by herself and with other people. Fish included eels, flounders, sharks and various types of tropical fish. Non-fish included turtles, whales and dolphins. In each instance the pigeon generalised its ability to respond more frequently to positive than to negative exemplars of a particular concept by performing appropriately to photographs it had not seen before.

It is still unclear just how the pigeon learns concepts. Unlike humans, who can label examplars of particular concepts, pigeons cannot. Even more puzzling is the pigeon's ability to form a concept in the absence of any contact with its examplars. Fish have not been a natural part of a pigeon's environment for millions of years.

The most reasonable explanation of a pigeon's ability to form a concept is that it comes equipped with analysers that can be trained to detect the critical features of certain classes of objects. We should remember, however, that whatever those features may be, we still do not know how to program the most sophisticated computer available to master the deceptively simple types of concepts pigeons can learn in a matter of days.

Descartes' view that animals cannot think was, in large part, based on his belief that animals lacked the capacity to engage in linguistic communication. While Descartes appears to have been right about the linguistic capacity of animals, recent studies of animal behaviour have shown clearly that animals do think. This state of affairs leaves us with a baffling but fundamental question that is likely to dominate the study of animal behaviour for the foreseeable future. How does an animal think without language?

The examples of animal thinking that I have described provide some tentative answers. Animals can form visual images. They can also form maps of their environments and produce representations that integrate and relate the elements of a sequence of images. Undoubtedly, we have only just begun to glimpse the nature of many long-neglected mental abilities of animals. Learning the nature of those abilities cannot help but provide an important biological benchmark against which to assess the evolution of human thought.

37

Can a bee behave intelligently?

JAMES L. GOULD and CAROL G. GOULD
14 April 1983

Most, but not all, of a bee's behaviour is the result of innate programming and places no intellectual demands on the bee. The few exceptions indicate that, like every other species including humans, a bee is just as clever as it needs to be.

When a foraging honey bee finds a new source of nectar or pollen, she returns to the hive to recruit help. She performs a ritualised dance that tells the other bees the distance, direction and quality of the food. They "memorise" the information she supplies, process it somehow and then, compensating for crosswinds and the movement of the sun, fly out on their own directly to the flower patch. This behaviour looks on the surface like a complex communication system, and the participants seem to be acting at least intelligently, even rationally. The more we learn about bees' capabilities, though, the more glaring their limitations become, and the question of what constitutes intelligence emerges as a central issue in understanding behaviour.

Largely as a result of Donald Griffin's book *The Question of Animal Awareness* (Rockefeller University Press, 2nd ed. 1981), ethologists have begun to re-examine the issue of animal intellect and to ask whether the organisms they study are, as we presume ourselves to be, something more than mere mindless circuitry. But the mind is, by nature, a private organ. How are we to judge from an animal's overt behaviour whether we are observing a well-oiled machine or a creature with some degree of intelligence and creativity? Particularly with insects, whose chitinous exoskeletons make it difficult to consider them in anthropomorphic terms, how are we to discover the extent to which they might be acting intelligently?

Several lines of evidence for insect intelligence have come to the fore, but a little careful thinking, observation and experimentation indicate that most of these criteria are untrustworthy. One intuitively powerful argument, for instance, is that since animals regularly face problems and solve them in sensible ways, they must have some intellectual grasp of the problem. When a honey bee, for instance, encounters a dead bee in the hive, it very properly tosses it out of the colony. But experience tells us that adaptive behaviour most often reflects the intelligence of evolution rather than that of the animals it has so carefully programmed. Bees recognise their dead colleagues by means of a "sign stimulus" – a single key feature of an object which is taken to represent the entire object. In this case a special "death odour", possibly oleic acid, releases the act of removal. So mindless is the wiring of this sensible hygienic behaviour that a drop of oleic acid on an otherwise innocuous piece of wood or even on a live bee results in the removal of the offending object. The sight of one bee carrying out a struggling sister or even the queen should convince us that behaviour can seem intelligent in its normal context without any need for the intellectual participation of the actors.

A second criterion frequently suggested is that the very regularity and invariability of such robot-like behaviour may be a guide to what behaviour is performed automatically and without the need for thinking. As we all know from personal experience, intellect will often come up with two very different solutions to the same problem in two different individuals, or even in the same individual on two different occasions. An automatic computer would come up with one "best" answer. When the 19th-century French naturalist Jean Henri Fabre interfered with the prey-capture ritual of a cricket-hunting wasp by moving its paralysed victim, he stumbled upon some of the wiring that runs the wasp's routine. The wasp, whose behaviour appears eccentric but intelligent, invariably leaves the cricket she has caught lying on its back, its antennae just touching the tunnel entrance, while she inspects her burrow. Each time Fabre moved it even slightly away from the entrance, the re-emerging wasp insisted on repositioning the cricket precisely, and inspecting the tunnel again. Fabre continued this trivial alteration 40 times and the wasp, locked in a behavioural "do-loop", never thought to skip an obviously pointless step in her programme. Clearly the wasp is a machine in this context, entirely inflexible in her behaviour.

The remarkable persistence of the wasp's performance serves also to remind us that most other animals have contingency plans to extricate them from such behavioural culs-de-sac. By far the most

K

common escape mechanism for organisms ranging from bacteria to human beings is "habituation", a kind of behavioural boredom by which an animal becomes less responsive as it encounters the same stimulus repeatedly. But habituation and other such escape strategies are not the result of any active intellect. They are merely sophisticated programming ploys, and in the sea slug *Aplysia* the neural and biochemical bases of the machinery are pretty well understood. The mindlessness of this acquired behavioural numbness is illustrated by the contrary phenomenon of sensitisation: almost any irrelevant but novel stimulus can instantly destroy habituation's insensitivity.

Other sorts of seemingly intelligent behavioural variability, though, cannot be accounted for either by "noise" in the computer or by habituation. Honey bees, for example, show spontaneous preferences for certain colours and shapes of artificial flowers, with many-petalled purple flowers being the most attractive to the apian mind. This display of aesthetic preference is not absolute, but probabilistic: given a choice between two colours that we know from learning experiments they can distinguish reliably – purple and blue, for example – the bees will choose their favourite, purple, only 70 per cent of the time rather than 100 per cent. Similarly, in a conflict an animal will sometimes fight and sometimes flee. Even in experiments in which care has been taken to factor out the role of immediate past experience, this sort of predictable variability persists.

Can the perplexing unreliability of animal behaviour be taken as evidence for something more (or perhaps less) than machinery making decisions? Probably not. Game theory demonstrates that it is usually most adaptive to be variable or unpredictable, so long as evolution or personal experience takes care to set the odds appropriately (see p. 171). Though flowers may more often be purple than blue, it makes sense to try blue-coloured objects from time to time rather than to concentrate exclusively on purple. Even this sort of quasi-aesthetic "decision" makes enough evolutionary sense that there is a good chance that it results from programming rather than intelligence, and in most carefully studied cases it is clear that variability *is* innate.

But though a great deal may be programmed into animals, there must surely be a limit to the complexity possible. There must be a point beyond which no set of built-in computer-like elements can suffice to account for an animal's apparent grasp of its situation, particularly in the face of variable or unpredictable environmental contingencies. The difficulty in drawing this intellectual line, however, is daunting. Some of the most impressively complex examples of behaviour we see are known to be wholly innate. The intricate knot-tying nest building of weaver birds is a case in point but, given the undoubtedly limited intellectual ability of the performer, surely the construction of orb webs is even more impressive. In total darkness, without prior experience, and with the location of potential anchor points for the support structure unpredictable, a mere spider sets about constructing a precise and complex network of several different kinds of threads held together with hundreds of

It takes no intelligence on the spider's part to build a complex and beautiful structure like this

precisely placed "welds". It even automatically repairs damage that occurs during construction. All this is accomplished through one master program and several subroutines and requires no conscious grasp of the problem.

The use of subroutines to deal with the unpredictable is especially obvious in navigation. Honey bees, for example, regularly use the Sun as their compass, compensating for its changing azimuth as it moves from east to west. This is a formidable task even for a human navigator but, as we have found out in the past few years, the bees' trigonometric adjustments are perfectly mindless, depending only on a memory of the Sun's azimuth relative to the bee's goal on the previous trip (or day) and an extrapolation of the Sun's current rate of azimuth movement. The strategy is innate, though the program must include steps for measuring the relevant variables when necessary. Bees recognize the sun by an equally innate criterion – its low ratio of ultraviolet (uv) to visible light – so that a dim, 10 degree, triangular, highly polarised, green object against a dark background is just as acceptable as the actual, intensely bright ½ degree, circular, unpolarised white Sun which the bees see almost every day in the normal sky.

When the Sun is invisible (obscured perhaps by a cloud, a landmark, or the horizon), the bees' whole Sun-centred system is discarded in favour of a backup system – a separate navigational subroutine – based on the patterns of polarised light generated in the sky by the scattering of sunlight. This analysis itself is composed of a primary and backup system, and uses sign stimuli and very simple processing. When the polarisation is unavailable as well (as on overcast days, for example), bees fall back on yet a third system, based on landmarks, and there is no reason to suppose we have exhausted the set of fail-safe plans built into bees. The apparent complexity of the formidable navigational behaviour that many insects display is, in fact, based on the interplay of groups of subroutines which are themselves quite simple. They depend as a rule on the same sorts of schematic stimulus–recognition systems and simple processing seen in less elaborate behaviour. Since a staggering degree of behavioural complexity can be generated by a set of individually simple subroutines, mere complexity of behaviour cannot be in itself a trustworthy guide to intelligence.

Another commonly accepted indication of intelligence is the way animals deal with the unpredictable contingencies of their world through learning, and it is here that our intuition tells us that we must be dealing with something very like intellect. After all, learning suggests to most of us some degree of understanding, some conscious comprehension of the problem to be solved. Alas, headless flies can learn to hold their legs in a particular position to avoid a shock, and even solve the problem faster than those still encumbered with brains.

Learning theory has traditionally recognised two general sorts of learning: associative learning (also known as classical or Pavlovian conditioning) and trial-and-error learning (also known as operant or Skinnerian conditioning). Associative learning is the process by which an animal comes to replace an innately recognised cue – the sign stimulus – with another cue or set of cues. It is nature's version of inductive reasoning: animals learn only cues which tend to *predict* the imminent arrival of something desirable, like food, and the reliability of the new cue does not need to be by any means perfect. Trial-and-error learning involves learning to perform a novel motor behaviour, which is used to solve a problem posed by nature. Animals discover by experimentation what works and what does not, and so experience shapes the behaviour. This process is in many ways analogous to deductive reasoning.

To understand the role of inductive and deductive learning in the lives of animals and how these processes relate to the issue of intelligence, let us look at how they work in honey bees. As we shall see, there is much in the organisation of bee learning that suggests the gears and wheels of an automatic pilot rather than any aware intelligence. When a honey bee discovers a flower, for example, she sets in motion a learning sequence which seems utterly mechanical in nature. A forager learns many things about a food source that aid her in the future, including its colour, shape, odour, location, nearby landmarks, time of nectar production, how to approach,

A bee that lands on a flower automatically learns all that it needs to make a second visit

land, enter, reach the nectar, and so on. Colour learning, for instance, has all the marks of associative learning: bees have an innate programme which recognises flowers by their dark centres and light petals (as seen in ultraviolet light – these markings are usually invisible to our eyes). After it has served its purpose, though, this sign stimulus is replaced by associative learning with a far more detailed picture of the flower. Bees learn colour only in the final three seconds as they land: the colour visible to the bee before the landing sequence, the colours it sees while standing on the flower to feed and while circling the blossom before flying off, simply never register. Experimenters can change them at will and the bee will never be fooled. A naive bee *carried* to the feeder from the hive and placed on the food source will circle repeatedly after taking on a load of sugar water as if "studying" the source, and yet when she returns a few minutes later she will be unable to choose the correct feeder colour. Yet, so mechanical is this learning routine that if we interrupt such a bee while she is feeding so she must take off and land again of her own accord, that landing permits her to choose the correct feeder colour on her next visit. Similarly, bees learn landmarks after taking off: a recruit who arrived and fed at the feeder, but was transported back to the hive while feeding, returned without the slightest memory of the landmarks she must certainly have seen on her arrival.

Other aspects of the associative component of flower learning seem equally curious. Although a bee learns a flower's odour almost perfectly in one visit, she must make several trips to learn its colour with precision and even then a bee never chooses the correct colour 100 per cent of the time. It learns shape less quickly, and time of day more slowly still. It is as though perfection, clearly possible in other contexts, is not in this case desirable. It appears, in fact, that the speed and reliability of a bee's flower memory at least roughly corresponds to the degree of variability it is likely to encounter among flowers of the same species in nature (including variation from day to day of an individual blossom). In fact, the rate at which a bee learns each component differs dramatically between various geographic races of honey bee, strongly implicating a genetic basis for the different learning curves.

Once a bee has learned how to recognise a particular kind of flower and when and where to find it, it is as though the information is stored in the manner of an appointment book. As a result, changing any component of the set – the odour, say, which is learned to virtual perfection after one visit – forces the bees to relearn painstakingly all the other pieces of information at their characteristic (slower) rates even though they have not changed. So, logical and impressive as the associative flower learning of honey bees seems, these hard-working insects appear simply to be well-programmed learning machines, attending only to the cues deemed salient by evolution (and then only in well-defined contexts and often during precise critical periods) – and then filing the information thus obtained in pre-existing arrays. Nothing in this behaviour, wonderful as it is, suggests any true flexibility or awareness. Nor is the situation any different when we look at the trial-and-error component of the behaviour by which bees learn how to harvest a flower species efficiently.

We can see that the widespread strategy of programmed learning is the means by which the genes tell their dim-witted couriers when and what to learn (how else could an insect reason it out?) and then what to do with the knowledge thus obtained. There are, however, cases of apparently self-directed learning that may admit of another explanation. Indeed, one of the main factors leading to the demise of classical behaviourism was the discovery that animals can learn a motor behaviour – which way to run in a maze to get to some food – without the need for either associative learning or overt trial-and-error experimentation. A rat, for instance, carried passively to each of two "goal boxes" at opposite ends of a runway and shown that one contains food and the other does not will, when released, run unerringly to the box with food. This phenomenon, which we call "cognitive trial and error", requires a deductive process to go on inside the mind of the animal without its actually *trying* different behaviours. The animal, be it the rat in its maze or a chimpanzee gazing from a group of boxes to a clump of bananas hung just out of reach overhead, must reason out a course of action in its *mind*. Here is something that seems very like intelligence and we must ask whether it is really that, or merely another clever but mechanical programming finesse that we do not yet see.

There are among honey bees three reported examples that appear at first glance to qualify as cognitive trial and error. One instance revolves around their avoidance of alfalfa (lucerne). These flowers possess spring-loaded anthers that give honey bees a rough blow when they enter. Although bumble bees (which evolved pollinating alfalfa) do not seem to mind, honey bees, once so treated, avoid alfalfa religiously. Placed in the middle of a field of alfalfa, foraging bees will fly tremendous distances to find alternative sources of food. Modern agricultural practices and the finite flight range of honey bees, however, often bring bees to a grim choice between foraging on alfalfa or starving.

In the face of potential starvation, honey bees finally begin foraging on alfalfa, but they learn to avoid being clubbed. Some bees come to recognise tripped from untripped flowers and frequent only the former, while others learn to chew a hole in the side of the flower so as to rob untripped blossoms without ever venturing inside. Who has analysed and solved this problem – evolution, or the bees themselves? It may be that both cases are standard, pre-wired backup ploys: differentiating tripped from untripped flowers could simply be a far more precise use of the associative learning programme, while chewing through may be a strategy normally held in reserve for robbing flowers too small to enter.

Another slightly eerie case is not so easy to dismiss. During training to find an artificial food source, there comes a point at which at least some of the bees begin to "catch on" that the experimenter is systematically moving the food farther and farther away. The pioneer of bee research, Karl von Frisch, recalls (and we have observed) instances in which the trained foragers began to *anticipate* subsequent moves and to wait for the feeder at the presumptive new location. This seems an impressive intellectual feat. It is not easy to imagine anything in the behaviour of natural flowers for which evolution could conceivably have needed to programme bees to anticipate regular changes in distance.

Along the same lines, we have on several occasions during experiments on bee navigation seen behaviour that appears to reflect an ability to form what experimental psychologists refer to as a "cognitive map". The classic example of this phenomenon is the ability of a rat in an eight-arm maze to explore each arm in a random order without inspecting any arm twice (see p. 267). This ability to form a mental map and then formulate behaviour (perhaps by imagining various alternative scenarios) seems very like the ability of chimpanzees to imagine the solution to the hanging-banana problem by stacking the boxes in their minds before actually performing the behaviour and, of course, the same process goes on in our own minds all the time.

The first hint of such an ability in bees came years ago when von Frisch discovered that bees that had flown an indirect route to a food source were nevertheless able to indicate by their famous communication dances the straight-line direction to the food. By itself, it is easy to interpret this ability as some sort of mindless, automatic exercise in trigonometry. Three years ago we trained

Honey bees communicate complex information about food sources to one another when they return from foraging

foragers along a lake and tricked them into dancing to indicate to potential recruit bees in the hive that the food was in the middle of the lake. Recruits refused to search for these food sources, even when we put a food source in a boat in the lake at the indicated spot. At first we thought that the foragers might simply be suffering from some sort of apian hydrophobia, but when we increased the distance of the feeding station so that the dances indicated the far side of the lake, recruits turned up in great numbers. Apparently they "knew" how wide the lake was and we were able to distinguish between sources allegedly in the lake and sources on the shore. We see no way to account for this behaviour on the basis of either associative or trial-and-error learning. This ability is accounted for most simply if we assume that the recruits have mental maps of the surroundings on which they somehow "place" the spots indicated by the dances.

This interpretation is further reinforced by another observation. While exploring the question of whether bees can use information about direction gathered on the flight back from the food, we transported foragers caught as they were leaving the hive for natural sources to an artificial feeder in the middle of a large car park hundreds of metres from the hive. After being allowed to feed, the majority of foragers circled the feeder and, in many cases, departed directly for the hive, which was out of sight. Many of the young bees, however, circled helplessly and never got home. When the successful foragers arrived at the hive, many danced to indicate the car park. Now, for a bee to know the location of a barren car park which had certainly not been on their list of flower sites, it seems most reasonable to suppose that they were able to "place" it on some sort of internal map and then work out the direction home. That only older (and presumably more experienced) bees were successful at this task is consistent with this interpretation.

Taking these cases at face value, does the apparent ability to make and use maps provide convincing evidence of active intelligence? And if so, why are bees so thoroughly mindless in other contexts? The second question is easier to speculate on than the first. Intuitively it seems reasonable to suppose that if we were designing an animal, we would "hard-wire" as much of the behaviour as possible. Where there is a best way of doing something, or finding out how to do something, there seems little point in forcing an animal into the time-consuming, error-prone, and potentially fatal route of trial-and-error learning. But where explicit programming will not serve, it seems equally reasonable to direct an organism to fall back on "thinking", particularly when the solution to a problem can then, as in the case of imprinting, be wired into the system for later service. Much of human behaviour seems to fall into this neurologically economical pattern: we work hard to master a problem, then turn the solution into a mindless, rote unit of behaviour. Difficult problems like learning to type, ride a bicycle, tie shoes, or knit seem almost impossible at first, but once learned become as matter of fact as breathing or walking.

But whether cognitive trial and error qualifies as intelligence is more difficult. On the one hand we can imagine how we might go about prewiring a Cartesian map, and how we could then encode the instructions by which the information to fill the map should be gathered, stored and used. On the other hand, there is increasing evidence that many of the intellectual feats of our own species – language acquisition, Aristotelian logic, categorisation, pattern recognition, and the like – are themselves based on pre-existing wiring and storage. The more we learn about the brain, the more clearly we see how its specialised wiring affects what we are. It may be that the question is one of degree: to what extent is a pocket calculator "intelligent"? Does a TI-59 with its hard-wired navigation module installed – a good approximation to a small part of a honey bee brain – qualify? What about a chess-playing machine, programmed to examine the board and then "imagine" thousands of possible moves and evaluate them in relation to each other? Or is it the provision for automatic self-programming such as we see when a flower trains a bee to exploit it that is intelligence?

The more we look at the behaviour of insects, birds, mammals and man, the more we see a continuum of complexity rather than any difference in kind that might separate the intellectual Valhalla of our species from the apparently mindless computations of insects. We see the same biochemical processes, the same use of sign stimuli and programmed learning, identical strategies of information processing and storage, the same potential for well-defined cognitive thinking, but very different storage and sorting capacities and, most of all, very different intellectual needs imposed by each species' niche. In short, the intelligence of insects, like that of our species, seems to be more than anything else the intelligence of evolutionary necessity.

38

Wolves in dogs' clothing

GAIL VINES

10 September 1981

The skill of the sheepdog in understanding and even anticipating the shepherd's commands seems remarkable and intelligent. But to a large extent the sheepdog is using instincts that derive from the hunting behaviour of its cousins the wolves.

Shepherds need their dogs; without them management of sheep would demand far too much labour, especially on Britain's extensive uplands. But how is this spectacular harmony of man and dog achieved? How are farmers able to develop this beast as such a cheap and reliable machine for herding sheep?

The answer lies at the heart of modern studies of animal behaviour and shows both how far we have come in understanding the behaviour of our fellow mammals and just how much we still have to discover. For in the skilled performance of the sheepdog we find a complex interaction between genetics and learning. The border collie, the most common sheepdog, has been selectively bred to fulfil its specialised role on the farm, and the breed is renowned throughout the world for its prowess. Out of the one thousand million sheep in the world, one third are kept in countries where the border collie is the chief working dog. There is a flourishing export trade for Britain in dogs descended from the champions of sheepdog trials. Yet farmers also stress the dog's need for careful training and years of experience with sheep in order to develop its powers to the full.

The working sheepdog has to carry out a great variety of tasks throughout the year. The dog's work varies somewhat from farm to farm but is most demanding on a large hill farm, where a strong, fast and free-moving dog is needed.

In the autumn the ewes need to be gathered before mating. A good hill dog makes a wide sweep of the boundary of the farm; once

A sheepdog's hunting instincts enable it to work with the shepherd to control even the most intransigent sheep

the sheep are moving they will join others. When the flock is collected the dog must bring it to pens where ewes are sorted into the groups that will be mated with various rams. The dog will also have to cope with the large, stroppy rams or "tups". It is hard work for the dog at "tupping", which lasts about five weeks, as each lot of ewes needs to be gathered daily on the hill around the ram that will run with them. After tupping, the dog herds the sheep to the top of the hill each day, so that they leave the bottom grazing alone and can come down to it as the weather gets worse. This work requires a "driving" dog, one that willingly and steadily presses sheep away from the shepherd, without frightening the sheep excessively, for such disturbances – leading to stress – can bring on nutritional diseases.

During winter the whole flock may be gathered to draw out small or thin sheep but generally the sheep are left alone. In winter storms the dogs are needed to find sheep lost in drifts; the dogs may point at a buried sheep much like a gun dog, although shepherds do not train the dogs to do this. Some dogs, perhaps those with a keener sense of smell, are better than others at this task.

Two weeks before lambing – which in the hills will be from the end of March until well into May – all ewes may be gathered for injections, and now the ewes need very gentle treatment. Lambing time is perhaps the busiest of all for the shepherd and his dog. Ewes and lambs must be constantly caught for one reason or another, and the dogs are invaluable for "mothering-on" – keeping a lost or orphaned lamb with a ewe until she accepts the lamb as her own.

In June the sheep are sheared and dipped in insecticide solution to protect them against parasites, notably the green-bottle fly, *Lucilia*, whose larvae can virtually eat a sheep alive. The dogs need to gather the sheep, herd them into pens, and keep a constant supply of sheep running through to the shepherd at work in the shearing pen.

Herding lambs at weaning – August to September – is another difficult task for dogs, since lambs are hard to control; they tend to scatter rather than clump together and are very skittish.

Finally, throughout the year the dog may be used to separate out a few sheep or a single one ("shedding" and "singling") so that the farmer can treat a diseased animal. How does the dog master these varied tasks?

Several of the behavioural traits that are important to the shepherd are apparently determined largely by breeding (genetics). For example, dogs vary in the amount of "eye" they show – "eye" being the tendency to freeze when sheep are immobile. Too much eye can be awkward when the shepherd is miles away and the dog is entranced by its quiet quarry. But dogs with some eye can control the sheep by merely staring at them with great concentration. Hardly surprisingly, the sheep seem to find this rather unnerving.

Also largely inherited is the dog's "style" – its tendency to lie down or "clap" when it stops. Some dogs tend to stand at such times. Barking and biting ("gripping") the sheep are less desirable characteristics which evidently may be in-bred, and are hard to repress in some dogs.

Finally, most sheepdogs tend from their earliest days not simply to chase sheep but to run around them so as to get the sheep between themselves and the shepherd. Puppies may try in this way to herd almost anything that moves, including geese or ducks or even a football. These sorts of behaviour appear to be quantitative traits under the control of many "additive" genes – genes working in concert. Such characteristics are not passed on from parent to offspring in such a clear-cut way as eye colour which is controlled by only one or a few genes. The genetic complexity of behavioural traits – good or bad – presents problems for the breeder, but farmers still have a strong regard for breeding and many dogs competing in trials today can trace their descent from the great champion Wiston Cap.

But whatever the importance of breeding, everyone agrees that early environment and training can make or break a dog. Young dogs are trained to respond to a set of about 10 signals, both audible and arm-semaphore. Farmers may use words or whistles and often have a unique set of signals for each dog (for instance, Welsh commands for one dog and English for another). The farmers use these signals to position the dogs appropriately and control their rate of movement.

The basic commands tell the dog where to move and when to stop. The universal "stop" command is a long single blast on a whistle. Farmers may also tell the dog to "lie down" without moving. The dog is asked to move to its left with "come by" and to its right with "away to me". "Come here" may be used to position a dog in a particular spot. Another command, "come on" (to your sheep), is the same whether the dog is bringing sheep towards the shepherd or driving them away – in either case it walks towards the flock. "Go back" means the dog should go away from the sheep, to find sheep out of sight. The farmer uses the command "steady" to slow down a dog who is pushing the sheep too fast.

The way the signals are given can also be used to modify the dog's speed or degrees of left and right; sharp whistles in quick succession tell the dog to speed up, and repeated "come by" or "away to me" commands can alter the tightness of the dog's circling. Finally, "that'll do" tells the dog that the task is done and the dog is to leave the sheep and return to the shepherd.

The dogs are trained to understand these signals through a variety of ingenious techniques. First, the dog is taught to lie down and to come to the shepherd's "that'll do", just as you might train any dog. Next, the dog is encouraged to "circle" the sheep. Most dogs do this in some fashion without training; the shepherd's task is to control this movement. The dog is sent out from behind the shepherd to the shepherd's right and told to lie down when it gets directly behind the sheep (see Figure 1), with the sheep between the shepherd and the dog. The dog is at a "12 o'clock" position, with the farmer at "6 o'clock". The farmer teaches the dog to "gather" by giving the command "come on" as he walks slowly backwards.

The dog is then taught to move clockwise by a cunning use of the dog's own behavioural predilections. With the dog at 12 o'clock the shepherd moves a little to his left. The sheep will then walk in the opposite direction, to the shepherd's right. The signal "come

Figure 1 *Training a sheepdog: a step-by-step guide.*

The outrun *The shepherd may move slightly to the right as the dog starts to run out to encourage the dog to run wide. He tells the dog to 'lie down' when it gets to the 12 o'clock position. A semi-circular outrun is desirable because the dog can get into position without disturbing the sheep*

Gathering *With the dog lying at the 12 o'clock position, the shepherd moves slowly backwards while telling the dog to 'come on'. If the sheep panic the dog is told to 'lie down'. Eventually the farmer stands farther from the sheep and remains stationary*

Moving clockwise *The shepherd moves to his left, causing the sheep to move slowly to his right. He then says 'come by' as the dog moves to the new 12 o'clock position (directly opposite the shepherd's new position). Eventually the farmer can stay in the same place; the dog moves on the command alone. Similarly the dog learns to move anticlockwise to the command 'away to me'*

Circling *The shepherd gets the dog in the 12 o'clock position and then moves to 2 o'clock and gives the command 'come by' repeatedly until the dog has run round the sheep to the other side (opposite the shepherd). Because the shepherd is standing near the sheep the dog learns to run wide around the sheep. Eventually the shepherd will give the command 'come by' while away from the sheep and the dog will circle them*

The drive *This is one of the hardest manoeuvres to teach. The shepherd first tells the dog to lie down behind the sheep in a walled lane (a), and then moves forward with the dog at his side (b). As the dog gains confidence the shepherd hangs back and encourages the dog to go forward in front (c and d), always stopping the dog if it attempts to flank the sheep*

by" (which means the dog should move to the shepherd's right) is given as the dog moves naturally to its left to the new 12 o'clock position, in order to "contain" the sheep. Eventually the farmer himself has no need to move; the dog will move clockwise to its left on the command "come by". During these movements the dog does not move closer to the sheep; it must move in a circular path, keeping the same distance from the sheep. A similar procedure is used to teach the dog to move to the shepherd's left (anticlockwise) to the "away to me" command.

The dog is next taught to maintain a wide "outrun" around the sheep. The shepherd stands by the side of the sheep and gives the "come by" command, so that the dog must run wide around both the shepherd and the sheep. With these commands the dog learns in time to circle the sheep even when the shepherd is farther away from them.

Next the dog is taught to drive sheep away from the shepherd. A hedged or walled lane is best for this. The shepherd stays behind the sheep with the dog at his side, and encourages the dog to move the sheep slowly up the lane. Eventually the shepherd hangs back a little and the dog finds himself pushing the sheep away on its own. The shepherd discourages the dog from "heading" the sheep – running around to the front of the group – by calling the dog back just as the shepherd feels the dog is about to attempt to circle the sheep.

The first step in training the dog to "go back" (to look for sheep out of sight) is to break its concentration on the sheep it is already working. This is done by splitting a small flock into two and commanding the dog to leave one flock and go to another.

The fascinating thing about all this training is the way the shepherd moulds the dog's natural behaviour. Domestic dogs were evidently derived from wolves some 8–10 thousand years ago. Certainly anyone who has seen wolves in zoos will have been struck by their dog-like appearance – only their heavier coat and long bushy tails readily distinguish them. In fact, recently biologists were unable to determine conclusively whether a large dog-like animal shot while killing a deer was a small timber wolf or a large Alsatian. A large heavily built guard and herd dog used by the Kurdish shepherds of Western Asia has a bone structure very similar to local wolves.

We might expect, then, that dogs will show some behavioural, as well as morphological, characteristics in common with wolves. So it might be worth looking at wolf hunting for insights into the way sheepdogs operate.

Wolves pursue and capture large animals, such as caribou, by hunting in packs. They show a high degree of cooperation in seeking and capturing their prey. The pack members, while searching, stay loosely together and when a prey is sighted they all become alert. The search phase sometimes ends with a kind of group ceremony – all the wolves mill around excitedly, touch noses, wag their tails, and then begin pursuit.

Hunting wolves are as disciplined and specialised as working sheepdogs

Most caribou can outrun wolves, so the wolves must select the most likely victim. They may test the herd by rushing it and if one animal falls behind the rest they concentrate their efforts on it. The appearance of a solitary prey animal and independent movement by a member of the herd seems to stimulate the wolves. On one occasion, a wolf was seen to adopt a hunting posture the moment a caribou moved away from its herd. The wolf moved between the caribou and the rest of the herd; the caribou immediately became alert and tried to rejoin the herd.

During the chase and capture a pack can be helpful in many ways; the wolves show a variety of cooperative strategies that may be vital to success and hence to survival. A caribou may lose its advantage of speed by running somewhat erratically, making sharp turns; the wolves then cut corners and gain on the caribou. A pack of six or seven wolves may be able to surround the caribou and force it to stop.

On one occasion when five wolves were following a small herd of caribou heading into a small clump of stunted spruce, one wolf hid directly in the path of the caribou while the other four circled the spruce and began a stealthy "drive" through it, pushing the caribou towards the hidden wolf. On another occasion a lone wolf chasing an adult caribou was seen to bite at the flanks of the caribou but was dislodged as the caribou swung round. A second wolf joined the first but the caribou still successfully charged the attacking wolves with its lowered antlers. Only when two more wolves arrived did they bring down their prey. A fleeing animal may circle and thus encounter other wolves resting or lagging behind.

A pack of wolves can separate young animals from defending adults; it would be most difficult for a single predator to do this. Some wolves were observed to chase a cow moose and hold her at bay while others killed the calf that she was trying to defend. One wolf can easily kill a sheep and sometimes even a deer or a caribou but even packs have a low success rate against healthy moose.

Cooperative hunting is made possible by the highly developed social behaviour of the animals and by their behavioural plasticity. Apparently, just as there is a great deal of morphological diversity in the wolf – a wolf can be tall and lean, or short and heavy, large or small – so there is a behavioural individuality which may lead to some sort of division of labour in a cooperating group. In experiments with captive wolves in semi-natural enclosures, some wolves made no attempt to capture introduced rabbits, and others made no captures but remained close to the pack during the hunt, and assisted in kills. A dominant male may often take the initiative in selecting prey.

Finally, wolves seem to have well-organised memories both of the location of resources and of the spatial relationships between them. A pack of wolves may range over 300 km², reaching most parts at least once every three weeks. The animals travel on established routes, but take short cuts and make detours over natural barriers and sometimes lead other members of the pack to kills. Finding prey is tricky, because large prey are clumped but sparsely distributed, and parent wolves must find their way back to the den once a day during late spring and summer to feed their young. In carrying out these tasks they travel in straight lines, suggesting that they have "destinations" and are not merely wandering.

The basic skills of the sheepdog dramatically reflect its wolfian ancestry. The sheepdog is very alert to any sheep that strays from the flock and will race off towards it, even without a command – a behaviour very similar to the wolf's immediate interest in a stray caribou. The dog also takes a great interest in "shedding" and "singling" – splitting off sheep from the flock – just as the wolf attempts to increase the vulnerability of its prey by separating it from the herd. The dog will then devotedly guard the isolated sheep, preventing it from rejoining the flock, as a lone wolf might keep its prey while waiting for reinforcements.

The whole stance of a dog as it holds sheep with its "eye" is very reminiscent of the hunting posture of the wolf, and of course relies for its effectiveness on the response of the sheep. The sheep must regard the dog as a potential threat and react by clumping together. Shepherds say that too much white on a dog is undesirable, apparently because the sheep are not very frightened of such a dog, perhaps because a white dog bears less resemblance to the ancestral wolf.

The tendency of sheepdogs to circle round the sheep and station themselves directly opposite the shepherd appears to be a direct behavioural carry-over from the cooperative hunting techniques of the ancestral wolves. In fact, if the dog does not show this tendency to "head its sheep" almost from its first contact with sheep, the shepherd will have great difficulty trying to teach the dog this behaviour.

Some dogs follow sheep more readily than gather

Domestic dogs, like wolves, are very social animals. When allowed to run wild, they will sometimes form stable packs. More commonly, the dogs are socialised from an early age with humans whom they come to regard as their social peers, or even social superiors. The concept of dominance hierarchies – which essentially envisages a social ladder with the pack leader at the top – is now known to be an oversimplification of complex social interactions in a natural social grouping of animals, yet some animals within a group do seem to have more control over resources important to the group. Evidently domestic dogs often come to regard a human as a dominant figure, in much the same way as their wild counterparts may regard the so-called "pack leader". Shepherds apparently make good use of this social response in dogs – farmers are very conscious of their role as pack leader to their dogs and attempt to foster it. The dogs continually look to see where the farmer is and if several are left uncommanded – while farmers confer, for example – the dogs will often gradually encircle the sheep and hold them contained, or gradually move them towards their shepherd. Shepherds remark that it is usually not a good idea to let a young untrained dog run with older dogs because the pup may come to regard one of the older dogs as pack leader and begin herding sheep towards that dog.

Most telling of all are Larry Shaffer's observations of dogs on a Cumbrian hill farm, recorded in the ITV film, "Man Bites Dog". He noticed that the tasks which farmers have difficulty in training dogs to perform were those that apparently do not occur in wild pack hunters. It is very difficult to train a dog to drive sheep *away* from a shepherd. The dog will initially always try to flank the sheep, to get round the other side of them and move them towards the shepherd. It is also difficult to get the dog to leave sheep it has gathered and to turn back and look for more. Again, this is understandable when the dog's behaviour is seen as a form of pack hunting.

Like wolves, sheepdogs vary greatly in their behaviour. Some will follow sheep more readily than gather; some dogs may be expert on the road but useless for gathering. Many farmers turn this individual variation to their advantage by having specialist dogs, using one for driving and another for gathering. The farmer also has to cope with dogs of vastly different temperaments; he uses a gentle tone of voice with the sensitive dog but is more stern with another.

Yet the shepherds exploit not only the fairly simple and discrete behavioural tendencies of the dog but also what might be called its "cognitive" abilities: its ability to learn about its environment and use this stored knowledge to guide its own behaviour. To succeed, a social carnivore such as the ancestral wolf needs to be able to predict the dynamic behaviour of its prey and to adjust its own behaviour in response to the prey and to other members of the pack. As a sheep herder, the sheepdog needs to know a lot about sheep and about its own effect on them.

It seems that through training, the dog comes to understand not only what it is supposed to do on hearing a command, but also what the goal is and what effects its own behaviour can have on the environment. The shepherd's commands become a symbolic language for the communication of intention. Farmers emphasise that initial lessons must be carefully stage-managed to give the dog confidence and to show it that it can control the movement of the flock through its own behaviour. Every effort is made to ensure that the dog will succeed. The shepherd uses a few very docile sheep during training and tries to prevent the sheep from bolting by moving slowly and stopping the dog if the sheep appear frightened. The dogs are not exposed to sheep until they are old enough to be able to outrun the sheep. When the shepherd is teaching a young dog to "go back" to look for more sheep, he makes sure that there are some sheep for the dog to find. An older dog can be sent out to check whether any are left or not, but a younger dog might become confused or lose confidence if it finds none.

Even traits which appear to have a large genetic component can be modified by experience which seems to confer an element of self-awareness to the dog. An experienced shepherd may cure a dog of excessive "eye" by bringing in another dog to take the sheep away; the eyeful dog – who might otherwise remain stationary as long as the sheep remains stationary – thus apparently learns that if it moves forward the sheep will move.

The behaviour of the skilled dog also shows signs of its understanding of the task at hand and the reaction of the sheep to its behaviour. At line-ups to the shearing pen one dog would always chase and bring back sheep escaping from the pens if they had not been sheared, but would leave the sheared sheep alone; yet it had not been trained to do this. Another could tell (presumably by smell) which ewes were near lambing and bring those sheep home uncommanded. A dog who had lost its sheep into a neighbour's flock through gathering them in one way tried on its own initiative a new route the next time.

Many experienced dogs will turn sheep into the correct field, getting into the right position early without waiting to be commanded. A dog can learn to pace sheep, automatically adjusting speed and distance from the sheep, and altering behaviour appropriately when faced with "wild" or "heavy" sheep. Dogs experience particular difficulty when faced with recalcitrant ewes with lambs. One such ewe which had split off from the main flock refused to be moved and faced the dog square on, stamping its hooves. The dog returned to the main flock, cut off several sheep, and brought them over to the stubborn ewe. The ewe promptly joined this group and the dog was able to move them all back to the main flock. By manoeuvring animals by anticipating their behaviour, sheepdogs appear to be acting as "managemental" ethologists. Such anecdotes are legion; experienced sheepdogs exhibit many feats of "initiative" while they work on vast hill farms, often at great distances from the shepherd.

Thus, sheepdogs present a unique opportunity to study the interactions of genetically-inspired behavioural tendencies and complex cognitive capabilities which are all involved in natural social behaviour such as hunting in packs. The triangle of man, dog and sheep provides an unusually accessible window into this behaviour as, long domesticated, dogs are perhaps unique among animals in both their eagerness and their ability to understand people. Detailed experimental studies of working sheepdogs are likely to give us many exciting insights into the newest area of research in animal behaviour, christened by Donald Griffin "cognitive ethology".

39

The many faces of animal suffering

MARIAN STAMP DAWKINS

20 November 1980

In order to treat animals humanely, we have to know what they experience as suffering. Putting ourselves in their place is a start, but such well-meaning anthropomorphism may be misplaced.

How do we know when an animal of another species is suffering? A difficult question, certainly, because "suffering" carries overtones of mental experiences like fear, pain or a generalised longing for freedom, which are difficult enough to gauge in our own species, let alone in others. But the question is fundamental to many of the current debates about animal welfare. Scientists and farmers are accused of using "torture" and "concentration camp methods". Their critics must clearly feel that they know how to judge when an animal is suffering. Defenders of intensive farms and of scientific experiments, on the other hand, often stand up for what they do on the grounds that their animals are not suffering. Difficult though it is to know about the mental experiences of other animals, many people are thus quite confident that they have found the answer. The trouble is that their answers turn out to be very different. Like the blind men and the elephant, people see a distorted picture of animal suffering because they only catch hold of one part of the truth.

Take "productivity". It is often argued that as long as farm animals are productive – laying plenty of eggs or growing fast – they cannot possibly be suffering. Unfortunately, productivity as it is often measured need have no connection at all with the welfare of individual animals. Productivity is about financial profit. It may actually be more profitable for a farm as a whole if a few individual animals are damaged or even die than to employ more men or invest in more comfortable cages. This is not to say that farm animals definitely do suffer, but in order to discover whether they do, one has to look much further than a balance sheet.

What, then, should we look at? Clearly, it is very important to know whether the animals are physically healthy. Disease and injury are generally acknowledged to be major causes of suffering. What is much more questionable is whether their presence or absence is enough to define the animal's state of health. Suppose we have established that the chimpanzees in a zoo or the hens in a battery farm look healthy and in good physical condition. We might have satisfied ourselves that, among other things, their coats or plumage were sleek and clean, their eyes looked bright and they were generally alert. We would then have to decide whether their confinement imposed mental suffering which did not affect their external condition. Several lines of evidence suggest that we should certainly not ignore other factors. For one thing, apparently healthy zoo and farm animals often show bizarre abnormal behaviour such as bobbing up and down or eating their faeces. An animal may also suffer intensely, but too transiently for any overt signs of injury to make themselves apparent. Transporting food animals, for instance, may take only a matter of hours. Even when this does not result in obvious weight loss or bruising, there may still be more subtle physiological consequences, such as changes in the ammonia content of muscles or in hormone levels.

The biggest difficulty is our inability to ask animals what they are feeling. With greater knowledge of the animals themselves, however, the lack of words may not turn out to be such a formidable barrier. The animal's behaviour may provide evidence enough. If we go about it in the right way, we may be able to interpret the animal's "vocabulary of suffering" even without words.

If we want to know whether a hen in a battery cage feels frustrated, we can discover how a hen expresses frustration by preventing it from achieving its goal. For example, the food dish of a hungry hen can be covered with a piece of clear perspex so that the bird can see the food but not get at it. Hens will peck vigorously at the perspex, but they also show a number of behaviour patterns which are taken to be characteristic of frustration. They preen themselves, but do so hurriedly and frantically; they peck at the floor and shake their heads rapidly from side to side. Equipped with this information, we can then look at hens in battery cages for similar signs. With this approach we allow the hens to tell us how they express frustration, rather than assuming that they must be

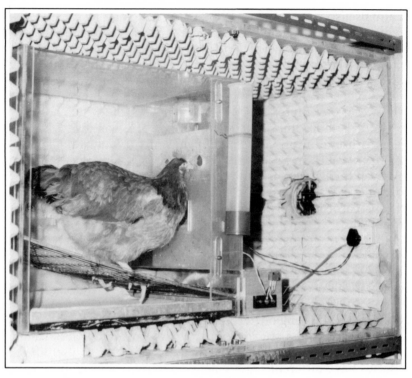

*This laboratory hen looks healthy enough, but is she happy on
that wire floor?*

frustrated in cages because we ourselves would be in similar
conditions.

There is an even more direct way of getting animals to express
what they feel by their behaviour. We can give animals access to
switches which control their environment in one way or another.
What happens when pigs are given the opportunity to adjust their
own levels of illumination? Work at the Agricultural Research
Council (ARC) unit at Babraham in Cambridge has shown that pigs
quickly learn what light switches are for and can be effectively asked
what sort of lighting they like and when they like the lights to go on
and off.

A variation on this theme is to give animals the chance to choose
between different environments and so to "vote with their feet" (see
New Scientist, vol. 80, p. 118). Hens that have been kept in battery
cages have shown clearly that they prefer to live in an outside run
rather than a cage. This is perhaps not particularly unexpected, but

animals also give surprisingly high priority to finding somewhere in which they can scratch and dustbathe. Although hens prefer spacious cages to small ones when there is no difference in flooring, they will choose to enter a tiny cage (so small that they can hardly turn round) if it has litter or sawdust on the floor. They will ignore cages many times bigger if they have wire floors. The hens can thus not merely show that they prefer litter floors to wire ones, but that the right sort of floor is more important than having space to move around in. This is useful information when we come to evaluate the relative merits of cages and deep litter from the hens' point of view. We can go even further and ask the hens how important flooring is in terms of some external yardstick, such as food. Is having a sawdust floor to dustbathe in "worth" going without food for different lengths of time? Thus we can get the hens to put a price on what they prefer.

Of course, animals do not always choose what is good for them, any more than people do. To rely on animal preferences only to evaluate their welfare would be as short-sighted as relying on productivity alone, or physical health or any other single measure. But such experiments offer an exciting and as yet almost untapped potential for finding out about animal feelings.

This brings out a crucial point. We have to *find out* about animal suffering, by careful observation or experiment. Because different species have different requirements, different lifestyles and, for all we know, different kinds of emotions, we cannot assume that we know about their suffering or well being without taking the trouble to study them species by species. The tapeworm's peptic Nirvana, as Julian Huxley once called it, is not for me, any more than my plodding air-breathing existence would suit a fish. It takes perhaps a certain amount of humility to recognise that species other than our own may suffer and feel pain. It takes even more humility to recognise that their subjective experiences may not be exactly or even remotely like our own.

To build up a picture of what an animal might be feeling, we need a great deal of factual information about its biology. Subjective experiences, by their very nature, cannot be studied directly in the same way that we might measure, say, the length of an animal's bones. But if we are prepared to take the trouble to collect it, we may be able to accumulate enough *indirect* evidence to tell us what we need to know. Even though we may be wrong, we are more likely to be near the truth if we study our animals first than if we rely on guesswork, however well-intentioned.

To be more specific, we need at least the following information

An experimental pig farm is said to be giving good results – but productivity does not necessarily reflect the wellbeing of the animals

about each species for which we are concerned, before we can claim to be able to recognise its signs of suffering. First, we need to document its signs of illness and health, including physiological disturbances which may be hidden from the unaided observer. Then we should try to find out how each species expresses its various emotional states. This may be quite a complex process, beginning with a study of how animals of that same species behave in the wild or relatively unrestricted state. In this way, we can obtain an idea of what the behavioural repertoire of the animal is and where the behaviour of captive animals differs from it. Chickens in battery cages, for instance, cannot roost or dustbathe in the way that less restricted fowl can. We should then go on to find out whether these behavioural restrictions lead to suffering, such as frustration, or whether they are merely more innocuous signs of captivity. Antelope in zoos or pigs in intensive farms may show less anti-predator behaviour than their wild counterparts, but it would be difficult to argue that they therefore suffered through being deprived of the opportunity to be chased by a predator. If, however, we discover that there are definite indications of prolonged fear, frustration or other signs, we would be led to argue that the animals are suffering. Either way, we would be able to reach a convincing conclusion only if we were thoroughly familiar with the ways in which that species expressed itself. We should also find out the animals' own preferences for different environments or conditions, as this might well alter our picture of how much they were suffering in a given instance.

Science is often thought of as having a rather negative and even sinister connection with animal welfare, causing suffering but not in any way alleviating it. It has, on the contrary, a major and positive contribution to make. By studying animals, we can learn how they express themselves and the conditions under which they suffer physically, and mentally. Through this knowledge, we have a much better chance of being able to ensure that they do not suffer while they are in our care.

Disagreement over whether an animal is suffering is not, clearly, the only reason why people argue about the welfare of animals on farms or in zoos and laboratories. Even if we could agree on this, there would still be contention over the relative importance of human well-being and animal suffering, the justification for certain sorts of experiments and so on. But if we did give more attention to the recognition of suffering in animals, we would go quite a long way to resolving some of the arguments. It would at least help us to avoid two of the most obvious, and opposite, dangers: seeing suffering where there is none and, worse, overlooking it because it does not have a human face.

40

Anthropomorphism: bad practice, honest prejudice?

D. R. CROCKER
16 July 1981

Ethologists are more ready to attribute human motives and emotions to their subjects in their popular presentations than in their scientific writing. And the popular accounts may come closer to what they really believe.

In the 1960s bookbuyers queued up to be reminded that underneath the fluff and sophistication of modern civilisation human beings are warty and beastly. The bestsellers containing these revelations were based mainly on ethology – studies of animal behaviour – and mostly on studies of apes in the wild. The writers had not always done the research themselves. Thus Desmond Morris, though a zoologist himself, was mostly reporting the ideas of others in his *Naked Ape* (Jonathan Cape, 1967), which has found its way into more than eight million homes; and Robert Ardrey, author of several books on the deep evolutionary roots of human behaviour, was a journalist. Ardrey's *The Territorial Imperative* (Collins, 1967) was subtitled "a personal inquiry into the animal origins of property and nations" and it argued that human beings were motivated to behave as they do by an innate urge to stake out territory.

The scientists, notably the animal behaviourists, whose studies inspired these and similar works were not quite sure how to take them. They were pleased that their ideas were so widely appreciated but also uneasy that these were not quite the ideas they meant. Liberal and left-wing scientists complained that ethology was being bent to the cause of conservatism – the idea that human behaviour had such deep biological roots suggested that it could not readily be changed simply by changing the human environment.

However, my researches of "popular" animal studies lead me to the impression that the popularisers, such as Morris and Ardrey, do not always pervert the meaning of the original work so much as exaggerate what is already there. More surprising than the "second hand" accounts of other people's work are the popular accounts that some ethologists have written of their own field studies. These are often far more subjective than most scientists usually care to be and, perhaps more to the point, they suffer from what has often been considered the unforgivable sin in biology. The sin is anthropomorphism, which the *Oxford English Dictionary* defines as "ascription of . . . a human attribute or personality to anything impersonal or irrational" – an "anything" which scientists conventionally take to include animals. But perhaps, by being subjective, they come closer to exposing what the scientist really thought. Perhaps these words should be seen not simply as popularisations, but as valuable complements to the "scientific" treatises.

Thus, for his doctorate at Oxford, John Mackinnon went to Borneo and Sumatra to watch orang-utans. As well as publishing a paper in *Animal Behaviour* (1974), Mackinnon told his story in a popular book called *In Search of the Red Ape* (Collins, 1974). In *Animal Behaviour*, he reports to his fellow scientists a brief "consortship" between two sub-adults (p. 56):

an adolescent female arrived . . . apparently by chance, at a fruit tree near which a sub-adult male had remained alone for two previous days. The animals could see each other, but not until the next morning did the male approach her. He climbed up into the same tree and fed close to her. She squealed and moved away each time he tried to reach out to her and she moved into another tree. The male followed, grabbed her and started to

Scholarship turned into best sellers

play with her in an old nest. She escaped from him and passed through another tree. He followed but remained in this next tree and performed a remarkable display . . . he again followed her and joined her in another long playabout. The female was still giving occasional squeals, but already seemed to have lost much of her fear and joined in the play of grappling, biting and tumbling about together in another old nest. The male made several attempts to hold and investigate the female's genital region, but no copulation was attempted. Later, the animals fed side by side and nested close to each other in the evening. The following day they broke up again.

In *In Search of the Red Ape* (p. 180), this becomes:

Towards evening . . . an adolescent female arrived to sample the *Dracontomelum*. Tom eyed her interestedly, but, surprisingly made no attempt to interfere with the trespasser. He maintained a polite distance while she dined but, when she was still there eating his precious fruit, the next morning, it was too much for him. As quietly as he could he stole up on the unsuspecting culprit and I thought he might cuff her soundly for her daring. Instead, he settled beside her with a friendly grunt and both animals started feeding.

This peaceful scene did not last long. Tom decided to speed things up and sidled across to better their acquaintance, but when he tried to stroke her she squealed in dismay and took to her heels. Undeterred, Tom swung after her and this time managed to get hold of his coy companion. He pulled her still squeaking into an old nest where the couple tumbled to and fro, biting and tickling playfully. Just for a moment Tom let go and his precious slipped away and crossed into the next tree. The young lady's bashfulness seemed to stir Tom to further displays of ardour and he paused in his pursuit to show her what he was made of . . . . It was a truly wonderful sight and seemed to have the desired effect on his intended. He hurried off after his new love, who ran just slowly enough to let him catch her again and soon they were rolling on another leafy couch, gurgling and wrestling. For the rest of the day they fed side by side in a mass of vines and that night they nested high in the same tree. I hoped this might be the start of a beautiful relationship, but somehow they did not quite hit it off and when I arrived at dawn Tom's shy girlfriend had given him the slip.

Here the animals are identified as individuals. They are named, their characters developed and their motives hinted at. A string of incidents is gathered into a drama – and they seem very human. Notice too, the censorship of the part where the male tries to inspect the female's sexual organs.

In his respectable book *The Mountain Gorilla* (University of Chicago Press, 1963), George Schaller, describing fighting among gorillas, informs us that (p. 291):

I have not witnessed serious aggressive contacts between gorillas. Although some animals appeared to quarrel violently on several occasions, the grappling, screaming and mock biting never resulted in a discernible injury. Only females appeared to be involved in these quarrels directly . . . . On at least three occasions several females not directly concerned with the quarrel entered into the scramble, a trait also noted in baboons . . . . Twice the dominant male stopped the squabbling merely by walking toward the animal and emitting annoyed grunts.

The popular rendition in *The Year of the Gorilla*, (Collins, 1964) runs as follows:

Since the gorillas are so closely associated day and night, tempers naturally become a little frayed at times, usually for trivial reasons. Quarrelling is usually confined to the females with the silverback male listening in aloof silence to the annoyed barks, which are husky and short like that of dogs. But bickering sometimes turns into a free-for-all, with females harshly screaming at each other in anger and grappling and biting. The male tolerates such a commotion only so long before he advances purposefully on the females. The screaming promptly subsides.

In both cases one's sympathy is directed toward the long-suffering male who must put up with these petty female disputes, but it is in the popular description that this interpretation is most openly paraded.

In her very popular book, *In the Shadow of Man* (Collins, 1971) Jane Goodall presents the following slice of chimpanzee life (p. 174):

Chimp in the shadow of Jane Goodall

Sometimes a male chimpanzee will actually insist on an unwilling female accompanying him on his travels until he is no longer interested in her, or she manages to escape. Indeed, this sort of masculine assertion of power led to a number of strange females being introduced to the feeding area in the old days . . . .

One day we shall never forget. Leakey, at that time, had become most peculiarly preoccupied with females. Constantly he forced one and then another to accompany him. On this occasion he had just lost his current victim and was sitting eating bananas in camp when Fifi arrived on the scene. She was pink and, instantly, his bananas forgotten, he stood up, his hair bristling, and shook branches at her. She ran quickly up to him and presented. He mated her and then the two of them sat grooming each other. Suddenly Leakey looked up and saw Olly approaching camp. At once all his hair stood on end again and he began waving branches at *her*. Olly hurried up and Leakey began to groom her. Fifi, looking innocent, walked very slowly away. But Leakey noticed, his hair rose once more, and Fifi ran back pant-grunting in submission. Then Leakey began to try to get both his females to follow, but neither wanted to go with him. First, he glared and shook branches at one until she ran up, and then at the other.

And so it went on until the strain and tension seemed suddenly to overwhelm Leakey – even as Fifi obediently approached him he ran at her and attacked her, rolling her over and over on the ground. At this Olly, of course, made a silent and rapid getaway and was soon lost to sight. The attack over, Leakey stood, his hair still on end, puffing from exertion, whilst Fifi, crouching to the ground, screamed and screamed.

When Leakey noticed that Olly had gone he moved rapidly some way up the slope, peered round, ran to the other side of the clearing and looked round from there. And so, of course, Fifi escaped. Not for 10 minutes or more did Leakey's hair slowly sink as he gave up the search and finally settled down to eat some bananas.

Readers of *Animal Behaviour* (1971, p. 161) get a much leaner offering:

On 10 occasions prior to 1965 mature males "forced" females to accompany them for periods up to three or four days. Only two of these females were receptive at the time and of these, one was pregnant, the other an adolescent and neither attracted much sexual attention from the other males . . . . On each occasion the behaviour of the male was similiar. He walked a few steps, looked back at the female, and, if she was not following, "branch-shook" at her. Normally she went after him; if not, he chased her, and (six times) this resulted in an attack. One female managed to escape – the male then searched for her for over 10 minutes, climbing trees, rushing up and down the slope where he had last seen her and peering in all directions. He did not find her and finally moved off.

Scientific journals require articles to be succinct and this may explain the relative terseness of the examples quoted here. But clearly the popular and academic expositions diverge in style. They are self-consciously written for different audiences. Sometimes this leads an author to draw nearly opposite conclusions from the same observations. Thus, in popular vein, Mackinnon's red apes commonly rape each other (*In Search of the Red Ape*, p. 176):

Ruby's face grimaced with fear as Humphrey seized her from behind and dragged her out of her nest. The ardent lover bit and struck her then, clasping his feet firmly around her waist, proceeded to rape the unfortunate female.

In *Animal Behaviour* Mackinnon repeats his view that "rape" is normal orang behaviour but he carefully muzzles the word between inverted commas. Moreover,

Due to the small size of the male penis and the difficulty of suspended copulation, it is probable that only when the female cooperates in the mating can successful intromission be achieved. In one observed instance of "rape" the female continued to struggle throughout and the male's penis could be seen thrusting on her back.

In academic translation, the inverted commas and qualification encourage us to put distance between animal behaviour and our own. *In Search of the Red Ape* gives its characters human names and its lurid style leaves us in no doubt that rape occurs in its full and horrifying human sense.

Keith Laidler also studied orang-utans for his PhD at Durham. He tried to teach a baby called Cody to speak English. After 14 months training, he told a scientific audience in *Action, Gesture, Symbol* (edited by Andrew Lock, Academic Press, 1978), p. 152:

Cody very rarely used his sounds to request some object presently out of sight, nor did he use one sound so as to put himself in a position to emit a second sound. This is considered due, in part, to the immaturity of the subject who was 15 months at termination of the experiment. Exceptions to this state were few, but did, nevertheless, occur, the protocol given below falling within the last week of the experiment and suggestive of a developmental, rather than an absolute, limitation on ability. Thus, on 15 October the infant twice refused the last of his pan-food voicing "kuh" each time. When placed on the floor, he immediately made his way to where the milk bottle was located. Later after termination of the experiment, but before Cody had been left to return to a more natural state with a second infant orang, the infant came across to the teacher when he offered pan-food, voicing two "Puhs". The teacher replied "No, fuh" on two separate occasions (the correct sound for pan-food) but the infant replied each time with a "puh", attempting at the same time to climb up the teacher. When allowed to do so, and settled on the teacher's right hip, he turned and without being prompted, uttered a good fuh-sound with his eyes directed towards the food.

In the version he wrote for a mass audience (*The Talking Ape*, Collins, 1980, p. 146) we find that:

As time progressed, our little "talking ape" began to use his simple functional language to insist on what he wanted even when it was out of sight. This began during Cody's 14th month of life during a feed session. Cody refused his last spoonful of cereal and instead uttered his "milk" word. I hesitated and Cody repeated "milk" in louder tones. I was intrigued as to what the young creature would do and placed him on the floor

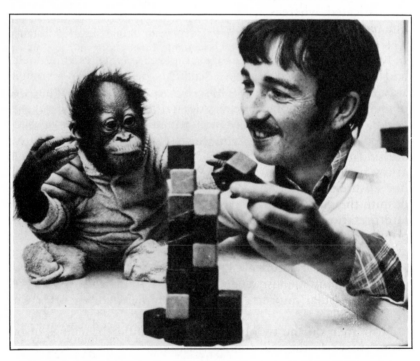

Keith Laidler with Cody, the orang-utan he attempted to teach to speak. His own estimation of his success depends on his audience

at my feet, whereupon Cody crawled to the kitchen for his cup of milk. Was he trying to tell me he wanted no more solid food while at the same time indicating what he preferred? Was he asking for his milk? I was not certain that he was, in fact, capable of such communicatory skill but a second incident convinced me.

Cody was playing on the floor when I approached with a pan of porridge and honey, a special treat the young orang really enjoyed. The following "conversation" ensued:

Cody: "Pick me up"

Me: (having expected Cody to utter his usual "food" sound) "No. What do you say?"

Cody: (halfway up my trunk) "Pick me up."

Cody then reached the upper portions of my chest, and settled himself comfortably on my right hip. Only then did he turn and looking straight at the plate of porridge, say "food".

In one version it is tentatively argued that Cody has some rudiments of speech which may possibly develop into something more sophisticated, whereas in *The Talking Ape* progress seems assured; one feels it is only a matter of time before Cody will be performing all sorts of linguistic tricks.

When I first came across these examples, I was rather smug. I felt I had caught so-called objective scientists pandering to a popular craving for animal stories whose heroes are hairy imitations of ourselves, Anthropomorphism is as old as Aesop's fables. It was odd, nevertheless, to find authors succumbing to it who had respectable scientific credentials, who ought to have known better, and indeed, had proved by their academic publications that they did know better. However, rereading the different accounts I am now not so sure. I did not find many cases of unabashed hypocrisy and, despite the euphemisms and speculation, it seems to me that the interpretation that carries the author's conviction is usually the popular one, the academic one coming across as cautious, but also as mealy-mouthed. I suspect the authors let their genuine feelings spill out into their nature books and that the academic pressure to be objective simultaneously encourages them to dissemble. My bet is that the popular informs the academic rather than the other way about.

How then do the two media call forth different tellings of the same story? Is the teller more trustworthy if he is impersonal? Surely, one style is not inherently more correct than the other? Rather, they appeal to different levels of awareness. The scientific style is reasoned and explicit, while the nature books are anecdotal, offering a hunk of personal experience which the reader is free to take or leave. Because the author as scientist is trying to persuade his peers of an argument, he necessarily gives them space to disagree. With the nature books we never engage our argumentative faculties but give ourselves up, instead, to the pleasurable effect. Whether in fact we find it pleasurable does not depend on our *agreeing* with the authors (we are being given a personal history, not a proposition), but on whether we are in tune with their commonsense. While the scientific mode invites us to listen critically, the nature books unveil a drama.

The difference in critical awareness demanded by the two genres is epitomised by the greater use, by academics, of inverted commas around risky words. Mackinnon, as we have seen, is careful to use them in front of other scientists, but when relating his experiences as wildlife adventures, he leaves words like "rape" unselfconsciously exposed. When writing for scientists he must excuse his use of the word. He is admitting that the concept of animal rape is a bit hard to swallow and is asking his readers to make allowances for the sake of the argument. Mackinnon the scientist needs to convince his audience and so must take care not to offend their sense of what is reasonable. The inverted commas instruct the reader to suspend criticism. But readers of *In Search of the Red Ape* did not enlist as critics. If Mackinnon the storyteller were to insert too many inverted commas he would be distinguishing between the animals and ourselves. The reader would then be obliged to consider the differences. Inverted commas, in advertising that "this is stretching a point", tend to disturb the smooth flow of the narrative and jolt the reader into scepticism.

I began by mentioning Morris and Ardrey. Their books, unlike the tales from nature, are explicitly propagandist. It is because these authors spoke "scientifically", I would argue, that they provoked the criticism they did. It was not what they said but the way they said it. The natural history shelves of our public libraries are teeming with accounts of wildlife behaving in most unnatural ways and no one complains (no one notices). But as soon as propagandists work anecdote into argument (that human beings are only animals and bound to compete for status, territory and so on) they are accused of bringing politics into science. In truth they are drawing out poorly hidden political prejudices already there.

Thus, although Morris and Co. at least argue their case (and so invite dissension), the scientific language they speak represses, rather than rises above, the original anthropomorphic impulse.

Self-critical ethologists have long known that fundamental categories of behaviour (such as aggression, dominance and territoriality) fetishised by the propagandist continually shift in meaning. Like magicians who challenge us to watch closely while they seemingly defy the laws of physics, the preposterousness of Morris's conclusions provokes some of us to look for sleight of hand. Often, however, we are fooled by magicians and we may similarly fail to find the flaw in the scientific argument, either because we secretly welcome its implications and do not probe too deeply, or because the logic is indeed impeccable. Illusionists succeed when we accept their props for what they physically appear to be. Propaganda likewise flaunts a superficial logic to draw attention from its fraught first principles.

The great advantage of the nature books is that, written off guard, their philosophical assumptions are on show for all to see. Instead of trying to wipe out subjectivity from the scientific literature, and succeeding only on toning it down enough to blend in with the surrounding facts, we ought to make it explicit. The tough-minded naturalist may wince to hear an animal described as looking "innocent" or "grief stricken". But I think it is valuable to know how the observers personally see their subject. In the first place, these phrases convey subtleties of meaning and context that are lost or suppressed when an interaction is reduced to a series of entries on a multichannel event recorder – the drama is unravelled as a string of incidents. In the second place, when anthropomorphism reconstitutes the richness of animal behaviour it also reveals the narrators and their biases. They cannot pretend to be neutral. When Jane Goodall tells me that "females are more likely than males to harbour grudges" (*In the Shadow of Man*, p. 122) or "she glanced at him with an expression that looked exactly like the smirk a little human girl might be expected to give under similar circumstances" (p. 123), I learn as much about Jane Goodall as I do about chimpanzees, and I bear it in mind.

In urging ethologists to take anthropomorphism seriously, I do not mean we should exalt it or relax scientific rigour. It is an imperative self-discipline to try to weed out what you saw happen from what you think it meant. The trouble with the complex social behaviour of primates, however, is that meaning is where the meat is. We cannot help humanising. Our understanding of ourselves and our societies is the material we inevitably use to build theories of animal social behaviour. We should not cover it up.

Conclusion

There is a sense in which understanding animals is beyond human capacity. We can never know what it is like to be an animal, just as we can never truly put ourselves in the place of another human being. Such metaphysical exercises are outside the scope of science. But on a more objective level, science has contributed and continues to contribute enormously to our understanding of others.

Scientific fashions have come and gone; ultimately there is no best route to follow. In the 1980s we can benefit from the insights as well as the mistakes of several generations of researchers. The modern ethologists' raw data, like those of the prewar behaviourists, are their observations of what the animal does – although they choose to make those observations under very different circumstances, and to subject them to different analyses. Only once you have a detailed and accurate record of your subject's activities as it goes about its daily business can you begin to ask the interesting questions whose answers will lead you towards understanding. The question "why" has many answers, all of which are right, and none of which is enough on its own. The causal answers, for instance, help us to understand the animal as an individual, the functional ones as a cog in a vast evolutionary machine. Both are equally valid, and there are yet others.

The 19th-century tendency towards anthropomorphism led to misunderstanding because it was not backed up with accurate observation. If today there is a move back towards attributing to animals characteristics hitherto thought exclusively human, it is because observations of animal behaviour have forced the conclusion that they do indeed possess these characteristics. The idea that there may be some sort of continuity in the mental abilities of humans and other animals has been out of favour since the turn of the century; no adequate evidence had been produced at that time. But some biologists are exploring this idea once more, influenced perhaps by the argument that since our physical evolution can be traced through lesser species, so our mental evolution should be.

In any case, it is no longer outrageous to suggest that some animals might themselves have something worth calling understanding. With the development of theories of animal behaviour that have some predictive power, we are at last in a position to claim an understanding of animals that is scientifically based. Good theory combined with computer techniques for data collection and analysis have invested the study of animal behaviour with a new rigour that fulfils the scientist's need for objectivity – the same need that gave rise to behaviourism. The hours spent watching animals in the wild now have a purpose greater than contributing to natural history. Sooner or later every sequence of observed behaviour will fit into a theoretical pattern that encompasses the entire animal kingdom. The extension of such theories to human behaviour is still contentious, but they enable us to speculate on the circumstances under which we first evolved from our ape-like ancestors.

At the moment the theoretical advances that illuminate the evolutionary significance of behaviour and the studies on animal "thought" that are more concerned with the mental life of the individual seem to have little to do with each other. But there is scope for another "new synthesis" in combining the two approaches. Once we have discovered the evolutionary criteria for the development of the sort of behavioural flexibility that can be called "intelligence", we may well be on the way to solving the problem at the heart of our own evolution.

Index